Universal Compression and Retrieval

Mathematics and Its Applications

Managing Editor:

M. HAZEWINKEL

Centre for Mathematics and Computer Science, Amsterdam, The Netherlands

Volume 274

Universal Compression and Retrieval

by

Rafail Krichevsky

Institute of Mathematics,
Russian Academy of Sciences,
Novosibirsk State University,
Novosibirsk, Russia

KLUWER ACADEMIC PUBLISHERS
DORDRECHT / BOSTON / LONDON

Library of Congress Cataloging-in-Publication Data

Krichevskiĭ, R. E. (Rafail Evseevich)
    Universal compression and retrieval / by Rafael Krichevsky.
        p.   cm. -- (Mathematics and its applications ; 274)
    Includes bibliographical references.

    1. Data compression (Computer science) 2. Data compression
(Telecommunication) 3. Information storage and retrieval systems.
I. Title. II. Series: Mathematics and its applications (Kluwer
Academic Publishers) ; v. 274.
QA76.9.D33K75   1993
005.74'6--dc20   93-45682

ISBN 978-90-481-4357-3

Published by Kluwer Academic Publishers,
P.O. Box 17, 3300 AA Dordrecht, The Netherlands.

Kluwer Academic Publishers incorporates
the publishing programmes of
D. Reidel, Martinus Nijhoff, Dr W. Junk and MTP Press.

Sold and distributed in the U.S.A. and Canada
by Kluwer Academic Publishers,
101 Philip Drive, Norwell, MA 02061, U.S.A.

In all other countries, sold and distributed
by Kluwer Academic Publishers Group,
P.O. Box 322, 3300 AH Dordrecht, The Netherlands.

Printed on acid-free paper

Contents

PREFACE

Objectives

Computer and communication practice relies on data compression and dictionary search methods. They lean on a rapidly developing theory. Its exposition from a new viewpoint is the purpose of the book. We start from the very beginning and finish with the latest achievements of the theory, some of them in print for the first time. The book is intended for serving as both a monograph and a self-contained textbook.

Information retrieval is the subject of the treatises by D. Knuth (1973) and K. Mehlhorn (1987). Data compression is the subject of source coding. It is a chapter of information theory. Its up-to-date state is presented in the books of Storer (1988), Lynch (1985), T. Bell et al. (1990).

The difference between them and the present book is as follows.

First. We include information retrieval into source coding instead of discussing it separately. Information-theoretic methods proved to be very effective in information search.

Second. For many years the target of the source coding theory was the estimation of the maximal degree of the data compression. This target is practically hit today. The sought degree is now known for most of the sources. We believe that the next target must be the estimation of the price of approaching that degree. So, we are concerned with trade-off between complexity and quality of coding.

Third. We pay special attention to universal families that contain a good compressing map for every source in a set.

There are some conventional sets, for instance, the set of Bernoulli sources, whose universal families consist of a sole map. Universal families are used, first, to encode unknown and changing sources and, second, to retrieve information optimally.

We present a new retrieval algorithm that is optimal with respect to both program length and running time. Its program length is $\log_2 e$ bits per letter. The running time equals the wordlength (in bitoperations) to within a constant factor. The time to precompute such an algorithm is polynomial in dictionary size. If the cardinality of the table encompassing a dictionary S exceeds $|S|^2$, then the program length becomes logarithmic in $|S|$.

New algorithms of hashing and adaptive on-line compressing are proposed.

All the main tools of source coding and data compression, like Shannon, Ziv-Lempel, Gilbert-Moore codes, Kolmogorov complexity, ε-entropy, lexicographic and digital search are exposed in the book.

We use data compression methods in developing
a) short programs for partially specified boolean functions
b) short formulas for threshold functions
c) identification keys

d) *stochastic algorithms for finding occurences of a word in a text*

e) *T-independent sets.*

The book may be considered a survey of Russian source coding. On the other hand, it relies on papers of P. Elias, A. Wyner, J. Ziv, M. Fredman, J. Kolmosz, E. Szemeredi, R. Karp and many others.

The book presents the theory of data compression which is a guide for practical algorithms.

Audience

The book is intended for information theorists, computer scientists and students in those areas. It may serve as a theoretical reference for communication engineers and data base designers. Some of its parts may be useful for biologists (in taxon making).

A precis of the content is presented in the introduction.

Acknowledgments

The book is based on the course of information theory, which was delivered many times at Novosibirsk State University. A considerable portion of the book concerns with the theorems of my students B. Ryabko, V. Trofimov, G. Khodak, L. Hasin and V. Potapov.

I am indebted to Natalie Martakov for professional typesetting. S. Kutateladze gave a lot of useful advice. The book is greatly influenced by my teacher, A.A. Lyapunov and by prominent Russian information theorists M.S. Pinsker and R.L. Dobrushin. I am grateful to professor Iorma Rissanen for support.

Introduction

(what the book is about)

Sources

There are two types of information sources. A combinatorial source produces a subset of words. A stochastic source can produce each word with a probability. We distinguish finite memory (Markov), zero memory (Bernoulli), stationary sources.

Entropy

Entropy of a finite combinatorial source is its log size. Entropy of a finite stochastic source is the average minus log probability of its letters. Entropy is defined for stationary sources.

Codes

We address injective coding maps, because they do not change the quantity of information carried by a message. Prefix codes, which constitute a subclass of injective ones, are especially interesting. Slight violation of injectivity is allowed sometimes. Nearly injective maps are called hashing functions. The degree of noninjectivity is measured by the colliding index.

Redundancy

The coding cost is the average quantity of output letters per input one. The redundancy is the excess of the cost over the entropy. The goal of data compression is to diminish redundancy. The less redundancy, the more coder and decoder complexity, which is characterized by coding delay, program length and running time.

Redundancy-Complexity Trade-off for Stochastic Sources

We discuss data compression and information retrieval from the standpoint of this trade-off. The redundancy of the Shannon, Huffman, Gilbert-Moore encodings of Bernoulli sources is reciprocal to the coding delay (blocklength). We prove that there are no source and no code with a better delay-redundancy relation.

The only exception is the Bernoulli source whose minus log probabilities of letters are integers. Per letter coding time of Markov sources (in bitoperations) is reciprocal to the redundancy to within a logarithmic factor.

1

There is a universal code, which compresses the output of any Markov source without any knowledge of its probabilities. It is interesting to note that the lack of knowledge deteriorates the redundancy-complexity relation moderately, by a logarithmic factor only.

Redundancy-Complexity Trade-off for Combinatorial Sources

Suppose we have a combinatorial source S (dictionary) and a coding map f giving a place in a table to each word of S. The redundancy is the difference between the log size of the table and the source entropy $\log |S|$. The non-dimensional redundancy ρ is the redundancy divided by the entropy.

The formula

$$\Pr f = |S|^{1-\rho+o(1)},$$

where $\Pr f$ is the programlength of a best coding map, expresses the redundancy-complexity trade-off for $\rho < 1$. If $\rho > 1$, then $\Pr f$ is logarithmic in the wordlength. So, $\rho = 1$ is a threshold. On its different sides the trade-off behaves quite differently. The running time of the coding map is minimal, the precomputing time is polynomial.

Universal Numerators

For each dictionary S of a given cardinality a universal numerator contains a map injective on S. Exhaustive search yields optimal numerators. A numerator corresponds to every method of information retrieval. Some of the numerators presented describe new methods of information search (fast lexicographic retrieval, two-step digital retrieval).

String Matching and Error-Correcting Codes

are among applications of universal numerators.

Hashing

We develop universal hash-sets of minimal size. They contain simple functions. A hash-function for a given dictionary can be selected from the set either at random or by greedy algorithm, which is fast enough.

Boolean Functions

Suppose a set of binary words (domain) is divided into two subsets. A boolean function indicates which subset a word belongs to. A function is total or partial depending on whether its domain is the set of all words of a given length.

Two-step digital retrieval yields a best algorithm to compute total functions. Optimal algorithms to compute partial functions are designed through the agency of the combinatorial source encoding.

Threshold boolean functions have two-fold relation to the information retrieval. First, threshold formulas are connected to the Kraft inequality. Second, such formulas are connected to universal numerators. That relation is benefical both ways, for threshold functions as well as for retrieval. Jablonskii class of boolean functions is invariant under substitutions of constant. It turns out to be a version of a stationary source. If such a class contains all most complicated functions, then it contains every boolean function.

Short Tables for Rapid Computation of Lipschitz Functions

We use a table to find a value of a function. The simplest table is just the collection of the values taken by a function at all points of its domain. That table is rather large, whereas the only one table read suffices to find a value. Using more complicated methods (interpolation etc) makes the table shorter, the running time – greater. Universal numerators give a possibility to construct for Lipschitz functions tables, which are asymptotically optimal by both its size and computing time.

Identifying Keys

Such keys are used to determine the name of a plant. The frequencies of plants are unknown. Universal codes help to diminish the quantity of attributes necessary for the identification.

Variable-to-Block Codes

Words of different length are encoded by words of the same length. It is often convenient.

Kolmogorov Complexity

A word may be given many different codes. Kolmogorov's code is absolute or universal in a sense. Any code may outperform Kolmogorov's only by a constant, which does not depend on the word encoded. Although uncomputable, Kolmogorov's code plays an important role in data compression as a lower bound of condensation.

NOMENCLATURE

$\log x$ $=$ $\log_2 x;\ l_2(x) = \log\log x.$

$l_k x$ $=$ $\log(l_{k-1}x);\ l_1(x) = \log_2 x.$

$|a|$ $-$ the length of a word a.

$|A|$ $-$ the cardinality of a set A.

$|a|$ $=$ $|a_1| + \ldots |a_k|$, if a is a vector, $a = (a_1, \ldots, a_k),\ k \geq 1.$

$a!$ $=$ $a_1! \cdot \ldots \cdot a_k!$, if $a = (a_1, \ldots, a_k).$

a^b $=$ $a_1^{b_1} \cdot \ldots \cdot a_k^{b_k}$, if a and b are k-dimensional vectors, $b = (b_1, \ldots, b_k).$

$o(x)$ $:$ $\lim \frac{o(x)}{x} = 0.$

$O(x)$ $:$ $\lim \frac{|O(x)|}{x} \leq \text{const}.$

$\text{Bin } x$ $:$ the binary notation of an integer x.

$\text{val } x$ $:$ the integer, whose binary notation is x.

$f \circ g$ $-$ the composition of maps f and g; $(f \circ g)(x) = f(g(x)).$

A^n $-$ the set of n-length words over an alphabet A, $n \geq 1.$

A^* $-$ the set of all words over A.

A^∞ $-$ the set of all sequences over A.

E^n $-$ the set of n-length binary words, i.e. the n-dimensional cube, $n \geq 1.$

$H(S)$ — the entropy of a combinatorial source S, where $S \subseteq A^n$; $H(S) = \log |S|$.

$H(S)$ — the entropy of a stochastic source S,

$$H(S) = -\sum_{i=1}^{k} p(A_i) \log p(A_i),$$

where $A_1, \ldots A_k$ are the letters generated by S.

$r_a(x)$ — the number of occurrences of a word a in a word x.

$CH_n(w)$ — the empirical combinatorial n-order entropy of a word w. It is the logarithm of the number of different n-length subwords of w.

$H_n(w)$ — the empirical n-order entropy of w,

$$H_n(w) = -\sum_{x \in A^n} \frac{r_x(w)}{|w|} \log \frac{n \, |r_x(w)|}{|w|}.$$

$\Delta(x)$ — the subtree of a tree Δ, whose root is x.

$L\Delta$ — the set of leaves of a tree Δ.

$|\Delta|$ — the quantity of nodes of a tree Δ.

$|L\Delta|$ — the quantity of leaves of a tree Δ.

$\text{Prec } x$ — the quantity of leaves (words) preceding a node (word) x in a tree or in a partition.

$E_\lambda f$ — the mathematical expectation of f with respect to the Dirichlet measure with parameter $\lambda = (\lambda_1, \ldots, \lambda_k)$, $k > 1$:

$$E_\lambda f = \frac{\Gamma(|\lambda|)}{\Gamma(\lambda_1) \ldots \Gamma(\lambda_k)} \cdot f(x_1, \ldots x_k) x_1^{\lambda_1 - 1} \ldots x_k^{\lambda_k - 1} dx_1 \ldots dx_k$$

$x_1 + \ldots x_k = 1$.

$i(f, S)$ — signature of a map $f : A \to B$ on a set S,

$$i(f, S) = (|f^{-1}(b) \cap S|), \quad b \in B$$

$I(f, S)$ –   *colliding index of a map $f : A \rightarrow B$ on a set S :*

$$I(f, S) = \frac{1}{|S|} \sum_b \left| f^{-1}(b) \cap S \right|^2 - 1.$$

$U_T(A, B, a)-$ *universal a-hash set. For each subset S, $S \subseteq A$, $|S| = T$ there is map*

$$f : \; A \rightarrow B, \quad f \in U_T(A, B, a), \quad I(f, S) \leq a$$

$U_T(A, B)$ –   *universal numerator. For each subset S, $S \subseteq A$, $|S| = T$, there is an injective map $f : \; A \rightarrow B$, $f \in U_T(A, B)$*

$V_T(n)$ –   *piercing set. For any partial boolean function f of n variables, $|\mathrm{dom}\ f| = T$, there is a boolean function $g \in V_T(n)$, $g(x) = f(x)$, $x \in \mathrm{dom}\ f$.*

Mem   –   *the membership function of a partition. Mem maps a word w to the atom containing w.*

$F(L, \Delta, C) -$ *the set of Lipschitz functions on an interval Δ with the constant L, $|f(0)| \leq C$, $|f(x_1) - f(x_2)| \leq L |x_1 - x_2|$, $x_1, x_2 \in \Delta$.*

$\alpha = \alpha(f, S)-$ *the loading factor of a map f on a dictionary S, $\alpha = \frac{|S|}{|B|}$, where B is the range of f, $|B|$ is the size of the table.*

$\mathrm{dom}\ f$ –   *the domain of a function f*

$\mathrm{Pr}\,(f)$ –   *the program complexity of a map P (in bits)*

$T(f)$ –   *the maximal running time of a program of f over all elements of $\mathrm{dom}\ f$.*

$\mathrm{PT}\,(f)$ –   *the precomputing time of f, i.e. the time required to construct a program of f.*

Itr   –   *interface map. Given a set and a family of maps, Itr is the number of a map, which in injective on the set.*

Rep   –   *representation function. Rep is injective on every atom of a partition A.*

$s(A)$ –   *the maximal size of atoms of a partition A.*

$\gamma(A)$ –   *the quantity of different sizes of atoms of A.*

CHAPTER 1

Information Source and Entropy

Entropy is the amount of information provided by an event.

Combinatorial source is a set; its entropy is the logarithm of its cardinality. Epsilon-entropy is the logarithm of cardinality of a minimal ε-net.

Stochastic source is a random variable; its minus entropy is the averaged logarithm of its probabilities. Stationary source sustains shifts. Its entropy is the limit of per letter entropies of its subsources. If the entropy of a stationary source is maximal, then it produces each word of an alphabet. A set of boolean functions sustaining substitutions of constants may be considered as a stationary source. The entropy of any set of boolean functions of common use is zero.

A word has got empirical entropies, both combinatorial and stochastic. They are determined by the frequencies of subwords of the word.

1.1 Finite Combinatorial Sources

A combinatorial source is a data generating machine. In order to be defined more formally such a source is identified with its output. Then we can say that a combinatorial source is a set of words over an alphabet. Considering English sentences as generated by a combinatorial source, we exclude all grammatically incorrect phrases. All grammatically correct ones are equally likely to appear. If it is not the case, i.e., some phrases are more probable than the others, then the concept of a stochastic source (Section 1.3) is more suitable.

A finite combinatorial source is a finite set S. Its main characteristic is entropy.

The idea of the quantity of information carried by an element of S is supposed to be intuitively clear. This quantity depends only on the cardinality $|S|$ of S, and the dependency is monotone. Two elements of S carry twice as much information as one element does, n elements - n times as much, $n \geq 2$. If $\varphi(|S|)$ is such a quantity, then φ is a monotone function and

$$\varphi(|S|^n) = n\varphi(|S|), \qquad n \geq 2.$$

Here $|S|^n$ is the number of all n-tuples (n-length words) constituted by elements of S.

Claim 1.1.1.
The only monotone function meeting the equality

$$\varphi(k^n) = n\varphi(k)$$

8

for any natural n and k is the logarithm:

$$\varphi(k) = C \log k.$$

C is a positive constant.

Proof.

Let a and b be natural, a, $b \geq 2$. For any $N > 0$ there is m such that

$$b^m \leq a^N \leq b^{m+1}$$

Using the monotonicity of φ yields

$$m\varphi(b) \leq N\varphi(a) \leq (m+1)\varphi(b) \tag{1.1.1}$$

The logarithm meets the same inequality

$$m \log b \leq N \log a \leq (m+1) \log b \tag{1.1.2}$$

From (1.1.1) and (1.1.2), we obtain

$$\left| \frac{\varphi(a)}{\varphi(b)} - \frac{\log a}{\log b} \right| < \frac{1}{N} \tag{1.1.3}$$

Inequality (1.1.3) holds for all N. Hence, for every a and b

$$\frac{\varphi(a)}{\varphi(b)} = \frac{\log a}{\log b}$$

and

$$\frac{\varphi(a)}{\log a} = \frac{\varphi(b)}{\log b}$$

$$\varphi(a) = C \log a, \qquad C = \text{const}$$

Q.E.D.

The quantity of information carried by an element of S is called the Hartley or combinatorial entropy of S. If the constant C in Claim 1.1.1 equals 1, then we obtain the binary entropy $H(S)$:

$$H(S) = \log_2 |S|$$

The entropy measure is called "bit". To identify an element of a two-element set, we need one bit of information.

A source S is often a set of binary n-length words, $n \geq 1$. Then the per letter entropy of S equals $h(S) = \frac{1}{n}H(S)$. A source is nontrivial, if its per letter entropy equals neither 0, nor 1. Suppose there is a sequence of combinatorial sources, whose per letter entropies have a limit. Such a sequence is called nontrivial, if that limit is neither 0, nor 1.

Let A be an alphabet, A^n, $n \geq 1$, be the set of all n-length words over A, w be a word, n be a divisor of $|w|$. The word w is the concatenation of $\frac{|w|}{n}$ words over the alphabet A^n. The logarithm of the number of different words among them, divided by n is called the empirical combinatorial entropy of w with respect to A^n and denoted by $CH_n(w)$. E.g., $CH_2(00001110) = \frac{\log 3}{2}$, $CH_4(00001110) = \frac{1}{4}$. The maximal number of different n-length words is $|A|^n$. Hence,

$$CH_n \leq \log |A| \qquad\qquad (1.1.4)$$

Siberian scientists Zh.Reznikova and B.Ryabko used the concept of the entropy in studying the language of ants.

The society of ants is well organized. It is subdivided into several groups. The members of each group play a special role in the life of the society. There are so called ants-explorers or scouts. Their duty is to find some food. There are also ants-feeders. Their duty is to carry the food discovered by the explorers to the nest. The scouts are much more clever than the feeders. There is a need in a communication line from explorers to feeders. To investigate this line a binary tree was made from matches joined with each other by plasticine. The tree has l levels, $1 \leq l \leq 6$. The number of its leaves (endpoints) is 2^l. The entropy of the set of leaves is $\log 2^l = l$. To identify a leaf, one can go to it starting from the root. An ant decides at each node whether to go left (0) or right (1). The l-length binary word of the decisions identifies the leaf.

The tree was put into a small basin with water. A nest with hungry ants was placed at the root. There were small containers with water at every leaf. The water in one of those containers was sweetened. A scout tasted the sweet water and came to the root over matches. It met the feeders waiting for it. There happened to be a kind of a conversation after which the feeders went straight to the sweet container via the shortest path. Hence, they were informed by the scout. The conversation time was found to equal $a + b \cdot l$, where a and b are constants, l is the binary entropy of the set of leaves. We can conclude that the scout says first: " there is some sugar". It takes it a minutes. Then it explains how to reach the sweet water of sugar, saying whether to go left or right at each node of the tree. It takes b minutes to communicate each bit of information. The information rate b varies from 0,7 to 1 bit/min. This experience clarifies the communicative ways of ants. More details are in Reznikova, Ryabko (1986).

Special attention should be paid to the combinatorial finite state sources (automaton sources). Such a source has a finite number k, $k \geq 1$, of states. Going from one state to another, it generates a letter of an alphabet A.

A matrix S corresponds to a source S. The element S_{ij} of S equals 1, if the source can go from the i-th state to the j-th one, else $S_{ij} = 0$, $1 \leq i, j \leq k$. A state is called initial, some states are called final. A word is said to be generated by a source, if all its letters are generated by it when going from the initial state to a final one. The

$(t + 1) \times (t + 1)$ - matrix S,

$$
S = \begin{pmatrix}
1 & 1 & 0 & \cdots & 0 \\
0 & 1 & 1 & \ddots & \vdots \\
0 & \ddots & \ddots & \ddots & 0 \\
\vdots & \ddots & 0 & 1 & 1 \\
0 & \cdots & 0 & 0 & 1
\end{pmatrix}
$$

corresponds to the source, which produces words with t units. The initial state is the first; the final one is $(t + 1)$-st. If the state of the source is not changed, it generates 1, in the other case it generates 0.

The Fibbonacci source generates words, in which there is at least one zero between every pair of units. Its matrix is $\begin{pmatrix} 1 & 1 \\ 1 & 0 \end{pmatrix}$; the first state is initial; both states are final. When remaining in the first state it outputs 0, when going from the first to the second or vice versa it outputs 1.

1.2 Epsilon-entropy

The entropy of a finite combinatorial source is the logarithm of its cardinality. The entropy of any infinite source equals infinity. We feel that an element of a square carries more information then an element of a segment. However, their entropies are equal. We are going next to introduce the concept of ε-entropy. It enables one to distinguish one infinite source (set) from another.

Let S be a subset of a metric space, ρ be the metric. A set S_ε is called an ε-net for S, if for any $x \in S$ there is $x_\varepsilon \in S_\varepsilon$ such that $\rho(x, x_\varepsilon) \leq \varepsilon$. A set, which for any $\varepsilon > 0$ has got a finite ε-net, is called totally bounded. A totally bounded and closed set is called a compact.

For instance, the interval $(0, 1)$ has a finite ε-net for any ε. Hence, it is totally bounded. On the other hand, it does not include its endpoints. Hence, it is not closed and is not compact. A line is closed but not totally bounded. Any finite dimensional cube or ball are compact. A set of functions

$$f, \; f : [0, 1] \rightarrow R, \; |f(0)| \leq 1, \; |f'(x)| \leq 1, \; 0 \leq x \leq 1,$$

is compact. The metric $\rho(f_1, f_2)$ equals $\max\limits_{0 \leq x \leq 1} |f_1(x) - f_2(x)|$.

Let S be a totally bounded set. The minimal entropy of its ε-nets is called its ε-entropy and denoted by $H_\varepsilon(S)$.

To lowerbound ε-entropy we define ε-capacity of a set S. A set is called ε-distinguishable, if the distance between any two its points is greater than ε. The maximal entropy of ε-distinguishable subsets of S is called ε-capacity of S. It is denoted by $h_\varepsilon(S)$.

Claim 1.2.1.
For any totally bounded set S and any $\varepsilon > 0$

$$h_{2\varepsilon}(S) \leq H_\varepsilon(S) \leq h_\varepsilon(S)$$

Proof.
Take a collections of balls, whose centers are points of a maximal ε-distinguishable subset of a set S. The set S will be covered by those balls due to the maximality. The right inequality is proven.

To prove the left inequality suppose that $H_\varepsilon(S) < h_{2\varepsilon}(S)$.

Take a collection of balls whose centers are the points of an ε-net. Then there should be at least two points of a 2ε-distinguishable subset in a ball. The distance between those points is not more than 2ε – a contradiction. Thus, the assumption $H_\varepsilon(S) < h_{2\varepsilon}(S)$ is wrong, and the left inequality is proven.

Let J_n stand for the n-dimensional cube, i.e.,

$$J_n = \{(x_1, \ldots, x_n) : \ 0 \leq x_i \leq 1, \ i = 1, \ldots, n\}.$$

Find the ε-entropy of J_n.

Theorem 1.2.1.   *The ε-entropy of the n-dimensional cube equals asymptotically* $n \log \frac{1}{\varepsilon}$, $\varepsilon > 0$:

$$H_\varepsilon(J_n) \sim n \log \frac{1}{\varepsilon}.$$

Proof.
Take an ε-lattice, i.e., the set $\left\{ (i_1\varepsilon, \ldots, i_n\varepsilon), 0 \leq i_k \leq \left\lfloor \frac{1}{\varepsilon} \right\rfloor \right\}$. Its ε-entropy is $n \left\lfloor \log \frac{1}{\varepsilon} \right\rfloor$ $\sim n \log \frac{1}{\varepsilon}$. For any $c > 0$ $\log \frac{1}{c\varepsilon} \sim \log \frac{1}{\varepsilon}$, $\varepsilon \to 0$. Thus, the ε-entropy does not change asymptotically when changing ε for $c\varepsilon$. The distance between any two points of the lattice is not less than ε. So, the lattice is an $\varepsilon/2$-distinguishable set and

$$h_{\varepsilon/2}(J_n) \geq n \log \frac{1}{\varepsilon}(1 + o(1)) \qquad (1.2.1)$$

On the other hand, the distance from a point $(x_1, \ldots, x_n)$ to the point $(\left\lfloor \frac{x_1}{\varepsilon} \right\rfloor \varepsilon, \ldots,$ $\left\lfloor \frac{x_n}{\varepsilon} \right\rfloor)$ does not exceed $\sqrt{n} \cdot \varepsilon$, consequently, the lattice is an $\sqrt{n} \cdot \varepsilon$-net for J_n and

$$H_\varepsilon(J_n) \leq n \log \frac{\sqrt{n}}{\varepsilon}(1 + o(1)) \leq n \log \frac{1}{\varepsilon} \cdot (1 + o(1)) \qquad (1.2.2)$$

Inequalities (1.2.1), (1.2.2) and Claim 1.2.1 yield the Theorem.

Next we will find the ε-entropy of a functional space. Let L, $|\Delta|$ and C be positive constants, $F(L, \Delta, C)$ be the set of functions $f : [0, |\Delta|] \to (-\infty, \infty)$, which meet the Lipschitz condition

$$|f(x_1) - f(x_2)| \leq L |x_1 - x_2|, \quad x_1, x_2 \in \Delta, \quad \Delta = [0, |\Delta|]$$

and which are bounded at 0:
$$|f(0)| \le C.$$
The distance is defined by $\rho(f_1, f_2) = \max_x |f_1(x) - f_2(x)|$.

Theorem 1.2.2. *The ε-entropy of the Lipschitz space $F(L, \Delta, C)$ equals asymptotically $\left\lceil \frac{|\Delta| L}{\varepsilon} \right\rceil$:*

$$H_\varepsilon(F(L, \Delta, C)) \sim \frac{|\Delta| L}{\varepsilon}$$

Proof.
Divide the segment Δ into $\frac{|\Delta| L}{\varepsilon}$ subsegments, whose length equals $\frac{\varepsilon}{L}$. The number $\frac{|\Delta| L}{\varepsilon}$ is supposed to be an integer, for simplicity sake. Let F' be the set of continuous functions f_ε, which are linear with the coefficients either L or $-L$ on each of subsegments. The value $f_\varepsilon(0)$ equals $k\varepsilon$, $k = 0, \pm 1, \dots, \frac{C}{\varepsilon}$.

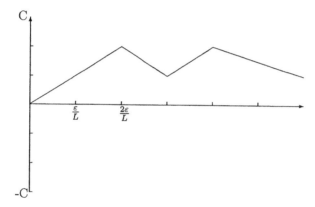

Fig. 1.2.1. A Lipschitz function

Evidently, $F' \subseteq F(L, \Delta, C)$.
The set
$$F'' = \{f_\varepsilon : \ f(0) = 0, \ f_\varepsilon \in F'\}$$
is a subset of F'. Each function of F'' is given a $\frac{\Delta L}{\varepsilon}$-length codeword, whose i-th letter is either $+1$ or -1, depending on whether the coefficient on the i-th subsegment is $+L$ or $-L$, $i = 1, \dots, \frac{|\Delta| L}{\varepsilon}$. The cardinality of F'' equals

$$|F''| = 2^{\frac{|\Delta| L}{\varepsilon}}$$

An element f_ε of F' is defined by both $f_\varepsilon(0)$ and an element of F''. For the cardinality of F' we get

$$|F'| = \frac{C}{\varepsilon} \cdot 2^{\frac{|\Delta| L}{\varepsilon}} \tag{1.2.3}$$

For any two functions $f_1, f_2 \in F''$ there is a subsegment, on which one of them has the coefficient $+L$ and the other $-L$. Hence, $\rho(f_1, f_2) \geq 2\varepsilon$, F'' is 2ε-distinguishable. From Claim 1.2.1 we obtain

$$H_\varepsilon(F(L, \Delta, C)) \geq h_{2\varepsilon} \geq \log |F''| = \frac{|\Delta| L}{\varepsilon} \tag{1.2.4}$$

Let $f \in F(L, \Delta, C)$. Next we will find a function $f_\varepsilon \in F'$ such that

$$f_\varepsilon(x) - 2\varepsilon \leq f(x) \leq f_\varepsilon(x), \quad x \in \Delta, \tag{1.2.5}$$

We will define f_ε inductively, proceeding from one subsegment to the other. Condition (1.2.5) is satisfied on the first subsegment for

$$f_\varepsilon(x) = \left\lceil \frac{f(0)}{\varepsilon} \right\rceil + Lx, \quad 0 \leq x \leq \frac{\varepsilon}{L}$$

Suppose it is satisfied on $1, \ldots, k$-th subsegments, $k \geq 1$. The function f meets (1.2.5) for $x = \frac{k\varepsilon}{L}$. It meets Lipschitz condition:

$$\left| f\left(\frac{(k+1)\varepsilon}{L}\right) - f\left(\frac{k\varepsilon}{L}\right) \right| \leq \varepsilon \tag{1.2.6}$$

As it follows from (1.2.5) and (1.2.6), either (1.2.7) or (1.2.8) holds:

$$f_\varepsilon\left(\frac{k\varepsilon}{L}\right) - \varepsilon \leq f\left(\frac{(k+1)\varepsilon}{L}\right) \leq f_\varepsilon\left(\frac{k\varepsilon}{L}\right) + \varepsilon \tag{1.2.7}$$

$$f_\varepsilon\left(\frac{k\varepsilon}{L}\right) - 3\varepsilon \leq f\left(\frac{(k+1)\varepsilon}{L}\right) \leq f_\varepsilon\left(\frac{k\varepsilon}{L}\right) - \varepsilon \tag{1.2.8}$$

On Fig. 1.2.2 the ordinate $P_1 A = f_\varepsilon\left(\frac{k\varepsilon}{L}\right)$, $AA_1 = 2\varepsilon$, $P_2 B = P_1 A + \varepsilon$, $BC = CC_1 = 2\varepsilon$. The value $f\left(\frac{k\varepsilon}{L}\right)$ belongs to AA_1 due to (1.2.5), $f\left(\frac{(k+1)\varepsilon}{L}\right)$ belongs to either BC, if (1.2.7) holds, or to CC_1, if (1.2.8) holds. In the first case the graph of f belongs to the parallelogram $A_1 ABC$. The line AB must be taken as the graph of f_ε on the $(k+1)$-th segment, on which the condition (1.2.5) will be met. In the second case the graph of f belongs to ACC_1A_1, the line AC is taken as the graph of f_ε. Thus, we can say that $f_\varepsilon \in F'$ is defined on $[0, \Delta]$ so as to meet (1.2.5). In other words, any $f \in F(L, \Delta, C)$, $f(0) = 0$, belongs to a set defined by (1.2.5). Any such set is an

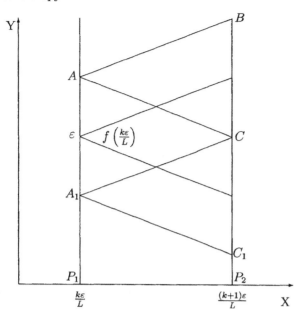

Fig. 1.2.2. Making an ε-net for the set of Lipschitz functions

2ε-diameter ball. The centers of those balls make an ε-net for the set of functions f, $f(0) = 0$, $f \in F(L, \Delta, C)$. Consequently, (1.2.3) and (1.2.4) yield

$$H_\varepsilon(F(L, \Delta, C)) \sim \frac{|\Delta| L}{\varepsilon},$$

Q.E.D.

The set F'' is an ε-net for $F(L, \Delta, 0)$. A function $g \in F''$ maps a point $\frac{k\varepsilon}{L}$, $k = 1, \ldots, \frac{|\Delta| L}{\varepsilon}$ to a point of the ε-net $i\varepsilon$, $i = 0, \pm 1, \ldots, \pm \frac{|\Delta| L}{\varepsilon}$ of the segment $[-|\Delta| L, |\Delta| L]$. Such a function is specified by a table with $\frac{|\Delta| L}{\varepsilon}$ locations. There is $\left\lfloor \log \frac{2|\Delta| L}{\varepsilon} \right\rfloor$ – length binary word $g(\frac{k\varepsilon}{L})$ at the k-th location. Collect the tables for all functions of F''. The set consists of $2^{\frac{|\Delta| L}{\varepsilon}}$ tables. Each table is given a $\frac{\Delta L}{\varepsilon}$-length binary number. The combined size of all the tables is

$$2^{\frac{|\Delta| L}{\varepsilon}} \cdot \frac{|\Delta| L}{\varepsilon} \cdot \log \frac{2 |\Delta| L}{\varepsilon} \quad \text{bits} \tag{1.2.9}$$

The set may be called universal: each function from the space $F(L, \Delta, 0)$ is ε-approximated by a piecewise linear function g. There is a table to compute g in the set. The computation is one table read, or $O\left(\log \frac{1}{\varepsilon}\right)$ bitoperations.

1.3 Stochastic Sources

Combinatorial source has been defined as a set. The elements of the set are equally likely to appear. If it is not the case, we will use the concept of stochastic source.

Finite stochastic source is a finite set S with a stochastic distribution on it,

$$S = \{A_1, \ldots, A_{|S|}\}, \quad P(a_i) \geq 0, \quad P(A_1) + \ldots + P(A_{|S|}) = 1.$$

The set S will often be called an alphabet; its elements will be called letters.

A partition A of a set S is a collection of disjoint subsets $A_1, \ldots, A_k$, whose union is S :

$$A_i A_j = \emptyset, \quad 1 \leq i, j \leq k, \quad \bigcup_{i=1}^{k} A_j = S.$$

The number k is called the cardinality of $A : k = |A|$. A subset A_i is called an atom of the partition A, $i = 1, \ldots, k$. Sets S, A_i, A can be infinite. There is an order on the set of all partitions of a set S. We say that a partition B follows a partition A, or B is more informative than A, if any atom of A is a union of some atoms of B. We use the symbol $\geq : B \geq A$. The minimal element in the set of all partitions is the partition, whose only atom is the set S itself. The maximal element is the partition, whose atoms are the elements of S. For any two partitions A and B there is a partition C, which is more informative than either A or B. The atoms of C are the intersections of the atoms of A and B :

$$C_k = A_i \cap B_j, \quad 1 \leq i \leq |A|, \quad 1 \leq j \leq |B|.$$

The Shannon entropy of a partition $A = \{A_1, \ldots, A_{|A|}\}$ of a set S is defined as

$$H(A) = -\sum_{i-1}^{|A|} P(A_i) \log P(A_i),$$

where $P(A_i)$ is the probability of the atom A_i, $0 \cdot \log 0 = 0$. The Hartley entropy of a partition is the logarithm of its cardinality:

$$H_{\text{Hartley}}(A) = \log |A|.$$

The entropy of a finite stochastic source is the Shannon entropy of its most informative partition.

If all elements of S have equal probabilities, then the Shannon entropy turns to the Hartley one. As it will be seen, the Shannon entropy is a good enough formalization of a vague intuitive notion "the quantity of information, carried by a letter".

Stochastic empirical entropy of a word w is defined analogously with the combinatorial one (Section 1.1). A word w is rewriten in the alphabet A^n, n is a divisor of

$|w|$. The symbol $r_x(w)$, where $x \in A^n$, is the number of occurences of x in w. The empirical per letter n-order entropy of w is

$$H_n(w) = -\sum_{x \in A^n} \frac{r_x(w)}{|w|} \log \frac{n\,|r_x(w)|}{|w|}.$$

For $w = 00001110$ we obtain $r_{00}(w) = 2$, $r_{11}(w) = 1$, $r_{10}(w) = 1$, $r_{01}(w) = 0$, $H_2(w) = -\frac{2}{8}\log\frac{2}{4} - \frac{1}{8}\log\frac{1}{4} - \frac{1}{8}\log\frac{1}{4} = \frac{3}{4}$.

Let A and B be two partitions of a set B. The partition A generates a partition on each atom B_j of B. The atoms of that partition are $A_1B_j, \ldots, A_{|A|}B_j$, $j = 1, \ldots, |B|$. The conditional probability of A_iB_j given B_j, equals

$$P(A_iB_j/B_j) = P(A_i/B_j) = P(A_iB_j)/P(B_j).$$

The entropy of the partition A on B_j is

$$H(A/B_j) = -\sum P(A_i/B_j) \log P(A_i/B_j)$$

The Shannon conditional entropy of A with respect to B is a weighted sum of $H(A/B_j)$:

$$H(A/B) = \sum P(B_j)H(A/B_j) = -\sum_j P(B_j)\sum_i P(A_i/B_j) \log P(A_i/B_j) =$$

$$= -\sum_{i,j} P(A_iB_j) \log P(A_i/B_j) \qquad (1.3.1)$$

Partitions A and B are independent, if

$$P(A_iB_j) = P(A_i)P(B_j), \quad 1 \le i \le |A|, \quad 1 \le j \le |B|.$$

Claim 1.3.1.
Let A and B be two partitions of a set. The conditional entropy of A with respect to B does not exceed the unconditional one:

$$H(A/B) \le H(A).$$

The inequality turns to the equality iff A and B are independent.

Proof.
We will use the Iensen inequality. If f is a convex function, $\lambda_i \ge 0$. $i = 1, \ldots, k$, $\lambda_1 + \ldots + \lambda_k = 1$, then for any $x_1, \ldots, x_k$

$$\sum_{i=1}^{k} \lambda_i f(x_i) \le f\left(\sum_{i=1}^{k} \lambda_i x_i\right)$$

The function f is convex, if its second derivative is nonpositive. The function

$$f(x) = -x \log x, \quad 0 < x < 1, \quad f(0) = f(1) = 0, \tag{1.3.2}$$

is convex.

The Iensen inequality for it turns to the equality if either all but one number λ_i equal zero, or all λ_i are equal, $i = 1, \ldots, k$. Apply the Iensen inequality to (1.3.1):

$$H(A/B) = \sum_i \sum_j P(B_j) f\left(P(A_i/B_j)\right) \le \sum_i f\left(\sum_j P(B_j) P(A_i/B_j)\right) = H(A) \tag{1.3.3}$$

The first Claim is proven.

The left side of (1.3.3) will be equal to the right one iff

a) the probability of an atom of B equals 1, the probabilities of all others equal to 0. In that case A and B are independent.

b) The probability $P(A_i/B_j)$ does not depend on j, but on i only, i.e., $P(A_i/B_j) = \varphi(i)$. In that case

$$P(A_i) = \sum_j P(B_j) P(A_i/B_j) = \varphi(i) \sum_j P(B_j) = P(A_i/B_j),$$

$$i = 1, \ldots, |A|, \qquad j = 1, \ldots, |B|.$$

The last equality means that A and B are independent, Q.E.D.

Claim 1.3.2.

The entropy is additive. It means that for any partitions A, B, C

$$H(A \vee B/C) = H(A/C) + H(B/A \vee C).$$

If C is empty, then

$$H(A \vee B) = H(A) + H(B/A).$$

Proof.

Let $A = \{A_1, \ldots, A_{|A|}\}$, $B = \{B_1, \ldots, B_{|B|}\}$, $C = \{C_1, \ldots, C_{|C|}\}$.

Obviously

$$P(A_i B_j/C_l) = P(A_i/C_l) P(B_j/A_i C_l).$$

We obtain from that equality, that

$$H(A \vee B/C) = -\sum_{i,j,l} P(A_i B_j C_l) \log P(A_i B_j/C_l) =$$

$$= -\sum_{i,j,l} P(A_i B_j C_l) \log P(A_i/C_l) - \sum_{i,j,l} P(A_i B_j C_l) \log P(B_j/A_i C_l) =$$

$$= -\sum_{i,l} P(A_i C_l) \log P(A_i/C_l) - \sum_{i,j,l} P(A_i B_j C_l) \log P(B_j/A_i C_l) =$$

$$= H(A/C) + H(B/AC).$$

The Claim is proven.

Claim 1.3.3.
The entropy is subadditive. For any partitions A and B

$$H(A \vee B) \leq H(A) + H(B).$$

The equality holds for both the Hartley and Shannon entropies.
Proof.
The inequality for the Shannon entropy is an immediate corollary of Claims 1.3.1 and 1.3.2. The quantity of atoms of the partition $A \vee B$ is not greater than the product of the quantities of atoms of A and B. It implies the inequality for the Hartley entropy, Q.E.D.

Claim 1.3.4.   (The entropy is a monotone function)
If $A \leq B$, then $H(B/C) \geq H(A/C)$. If C is empty, then $H(B) \geq H(A)$.
Proof.
From $A \leq B$ we get $A \vee B = B$. Then, from Claim 1.3.2

$$H(A \vee B/C) = H(B/C) = H(A/B \vee C) = H(A/C) + H(B/A \vee C) =$$

$$= H(A/C) + H(B/A \vee C).$$

Take into account that
$$H(A/B \vee C) \geq 0.$$

We get $H(A/C)$, Q.E.D.

Claim 1.3.5.
Conditional entropy is a monotone function. If $C \geq B$, then

$$H(A/C) \leq H(A/B).$$

It means that the more informed is an observer, the harder it is to surprise him (or her).
Proof.
Write the Iensen inequality for function (1.3.2):

$$\sum_l P(C_l/B_j) f\left(P(A_i/C_l)\right) \leq f \sum_l P(C_l/B_j) P(A_i/C_l), \tag{1.3.4}$$

$$1 \leq i \leq |A|, \qquad 1 \leq j \leq |B|.$$

By the condition, $B \leq C$. Hence, each atom of B is the union of some atoms of C, and for any i, $1 \leq i \leq |A|$, j, $1 \leq j \leq |B|$,

$$\sum_i P(A_i/C_l) P(C_l/B_j) = \sum_{C_l, C_l \leq B_j} \frac{P(A_i C_l)}{P(C_l)} \frac{P(C_l)}{P(B_j)} = \frac{P(A_i B_j)}{P(B_j)} = P(A_i/B_j). \tag{1.3.5}$$

Multiply (1.3.4) by $P(B_j)$ and sum over $1 \leq i \leq |A|$, $1 \leq j \leq |B|$. Take into account (1.3.5). We get the Claim.

1.4 Stationary Stochastic Sources

Let A be a finite alphabet, A^∞ be the set of infinite words over A. There is a stationary, i.e., independent on any shift of time, measure on A^∞. The set A^∞ with a stationary probabilistic measure on it is called a stationary stochastic source.

We get a Bernoulli source, if the letters of A are independent. The probability of a word w equals

$$p(A_1)^{r_1(w)} \ldots p(A_{|A|})^{r_{|A|}(w)},$$

where $p(A_i)$ is the probability of the letter A_i, $1 \leq i \leq |A|$, $r_i(w)$ is the number of occurences of A_i in w. We get a first order Markov source, if the probability of a letter depends on the preceding letter only. Such a source is described by a matrix $P = (p_{ij})$, $1 \leq i \leq |A|$, where p_{ij} is the probability to meet the letter A_j after the letter A_i. A vector

$$p = (p_1, \ldots, p_{|A|}), \qquad Pp = p,$$

is called stationary. It exists for any Markov source. The probability of a word $A_{i_1} A_{i_2} \ldots A_{i_k}$ equals $p_{i_1} p_{i_1 i_2} \ldots p_{i_{k-1} i_k}$.

We get a t-order Markov source, $t > 0$, if the probability of a letter depends on the preceding t letters.

The symbol S_i^j stands for the set of all generated by S words, whose j-th letter is A_i, $1 \leq i \leq |A|$, $j = 1, 2, \ldots$. The sets S_i^j, $i = 1, \ldots, |A|$, constitute a partition of S. The symbol S^j stands for that partition:

$$S^j = \left\{ S_1^j, \ldots, S_{|A|}^j \right\}$$

Define the entropy $H(S^n / S^{n-1} \vee \ldots \vee S^1)$ of a letter under the condition that n preceding letters are known, $n = 2, 3, \ldots$. The source S being stationary, the probabilities of its words do not change with any time shift. That is why

$$H(S^n / S^{n-1} \vee \ldots \vee S^2) = H(S^{n-1} / S^{n-2} \vee \ldots \vee S^1) \qquad (1.4.1)$$

Claim 1.4.1.
For any stationary source the sequence of conditional entropies $H(S^n / S^{n-1} \vee \ldots \vee S^1)$ goes to a limit monotonically, as n goes to infinity. The source may be either combinatorial or stochastic.

Proof.
Conditional entropy is monotonic (Claim 1.3.4). The partition $S^{n-1} \vee \ldots \vee S^1$ majorizes the partition $S^{n-1} \vee \ldots \vee S^2$. Hence

$$H(S^n / S^{n-1} \vee \ldots \vee S^1) \leq H(S^n / S^{n-1} \vee \ldots \vee S^2). \qquad (1.4.2)$$

We get from (1.4.1) and (1.4.2) that

$$H(S^n/S^{n-1} \vee \ldots \vee S^1) \leq H(S^{n-1}/S^{n-2} \vee \ldots \vee S^1),$$

i.e., the sequence $H(S^n/S^{n-1} \vee \ldots \vee S^1)$, $n = 2, 3, \ldots$ is monotonically decreasing. Thus, the sequence has a limit, Q.E.D.

Claim 1.4.2.
For a stationary source S the sequence $\frac{1}{n}H(S^1 \vee \ldots \vee S^n)$ goes monotonically to a limit, which equals the limit of conditional entropies:

$$\lim_{n \to \infty} \frac{1}{n} H(S^1 \vee \ldots \vee S^n) = \lim_{n \to \infty} H(S^n/S^{n-1} \vee \ldots \vee S^1).$$

The value of those limits is called the entropy of S and denoted by $H(S)$.

Proof.
Use the additivity of the entropy (Claim 1.3.2):

$$H(S^1 \vee \ldots \vee S^n) = H(S^1) + \ldots + H(S^n/S^{n-1} \vee \ldots \vee S^1). \tag{1.4.3}$$

We obtain from that:

$$\frac{1}{n}H(S^1 \vee \ldots \vee S^n) - \frac{1}{n+1}H(S^1 \vee \ldots \vee S^{n+1}) =$$

$$= \frac{1}{n(n+1)} \left(H(S^1) + \ldots + H(S^n/S^{n-1} \vee \ldots \vee S^1) - nH(S^{n+1}/S^n \vee \ldots \vee S^1) \right).$$

$$\tag{1.4.4}$$

From (1.4.3) and Claim 1.4.2 we can see that the sequence $\frac{1}{n}H(S^1 \vee \ldots \vee S^n)$ is monotonic. It is the arithmetic mean of the sequence $H(S^n/S^{n-1} \vee \ldots \vee S^1)$. Thus, by a known theorem, the limits of those sequences are equal, Q.E.D.

For a Bernoulli source any partition S^n does not depend on the partitions S^{n-1}, $S^{n-2}, \ldots, S^1$, $n = 2, \ldots$. It yields for the entropy $H(S)$:

$$H(S) = H(S^n/S^{n-1} \vee \ldots \vee S^1) = - \sum_{i=1}^{|A|} p(A_i) \log p(A_i).$$

For a Markov first order source any partition S^n depends on S^{n-1}, but not on S^{n-2}, $n = 3, \ldots$. We obtain

$$H(S^n/S^{n-1} \vee \ldots \vee S^1) = H(S^n/S^{n-1}) = H(S^2/S^1) =$$

$$= H(S) = \sum_{j=1}^{|A|} p_j H(p_{j1}, \ldots, p_{jk}).$$

The quantity $H(p_{j1}, \ldots, p_{jk})$ is the conditional entropy under the condition that j-th letter appears, $j = 1, \ldots, k$. The entropy $H(S)$ is a weighted sum of those entropies. The weights are equal to the stationary probabilities.

Claim 1.4.3.
For any Markov t-order source S, $t \geq 1$, there is a constant $C > 0$ such that for every n, $n \geq 1$

$$H(S^1 \vee \ldots \vee S^n) \geq nH(S) + C.$$

Proof.
Let t be the order of the source, $t > 0$. Take the sequence

$$H(S^1) \geq H(S^2/S^1) \geq H(S^3/S^2 \vee S^1) \geq \ldots$$

$$\geq H(S^{t+1}/S^t \vee \ldots \vee S^1) = H(S).$$

If $H(S^1) > H(S)$, then, by (1.4.3)

$$H(S^1 \vee \ldots \vee S^m) > \big(H(S^1) - H(S)\big) + nH(S).$$

The Claim is proven in that case with $C = H(S^1) - H(S)$.
If $H(S^1) = H(S)$, then $H(S^1) = H(S^2/S^1)$.
According to Claim 1.3.1, S^2 and S^1 are independent. Analogously, S^3 does not depend on $S^1 \vee S^2$, $\ldots$, S^{t+1} does not depend on $S^1 \vee \ldots \vee S^t$. Hence, any partition does not depend on any other, and S is a Bernoulli source in that case. Q.E.D.

1.5 Stationary Combinatorial Sources and Jablonskii Invariant Classes

Let A be a finite alphabet, A^∞ be the set of infinite words over A, A^* be the set of finite words over A. We will define two kinds of combinatorial stationary sources.

Let $S \subseteq A^\infty$. Take an infinite word from S and cross its first letter out. If the word obtained remains an element of S, then S is called a stationary combinatorial source. For instance, the set of all words generated by a finite automaton, whose every state is initial, is such a source.

Let $S \subseteq A^*$. Take a finite word from S and cross either its first or last letter out. If the words obtained remain elements of S, then S is called a stationary combinatorial source. It generates finite words.

Let $S \subseteq A^\infty$, the entropy be understood in the Hartley sense,

$$h(l) = H(S^1 \vee \ldots \vee S^l), \quad l \geq 1.$$

By the definition (Section 1.3), $h(l)$ equals the logarithm of the number of different words generated by S at the moments $1, \ldots, l$. The source S being stationary,

$$H(S^i \vee \ldots \vee S^{i+l}) = H(S^1 \vee \ldots \vee S^{l+1}), \quad i \geq 1. \tag{1.5.1}$$

If a and b are natural, then the subadditivity of the Hartley entropy (Claim (1.3.3)) and (1.5.1) yield:

$$h(a + b) = H(S^1 \vee \ldots \vee S^{a+b}) \leq$$

$$\leq H(S^1 \vee \ldots \vee S^a) + H(S^{a+1} \vee \ldots \vee S^{a+b}) = h(a) + h(b) \qquad (1.5.2)$$

If $S \subseteq A^*$, then let $h(l)$ be the logarithm of the number of l-length words, which belong to S. Any $(a + b)$-length word of S consists of an a-length beginning (prefix) and an b-length ending. By the definition of the stationarity, both belong to S. Thus,

$$h(a + b) \leq h(a) + h(b),$$

and inequality (1.5.2) holds for the both kinds of sources.

We reproduce the short proof of a Claim of M. Fekete. It was proven in 1923.

Claim 1.5.1. (M. Fekete)
Let h be a nonnegative function, $h(a + b) \leq h(a) + h(b)$, a, b be natural. Then there exists $\lim\limits_{n \to \infty} \frac{h(n)}{n}$, and $\lim\limits_{n \to \infty} \frac{h(n)}{n} = \inf\limits_{n} \frac{h(n)}{n}$.
Proof.
Let

$$\inf_{n} \frac{h(n)}{n} = \alpha \geq 0.$$

There is a natural number m such that

$$\frac{h(m)}{m} < \alpha + \varepsilon, \quad \varepsilon > 0. \qquad (1.5.3)$$

Take a natural n and divide it by m. We obtain

$$n = qm + r, \qquad 0 \leq r \leq m - 1 \qquad (1.5.4)$$

The condition of the Claim and (1.5.4) yield

$$h(n) \leq qh(m) + h(r). \qquad (1.5.5)$$

From (1.5.4) and (1.5.5) we get

$$\frac{h(n)}{n} \leq \frac{h(m)}{m} \frac{qm}{qm + r} + \frac{h(r)}{n}. \qquad (1.5.6)$$

If n goes to infinity, so does m, r and $h(r)$ remain bounded. If n is large enough, then

$$\frac{qm}{qm + r} < 1 + \varepsilon, \qquad \frac{h(r)}{n} < \varepsilon. \qquad (1.5.7)$$

Inequalities (1.5.3) and (1.5.7) yield

$$\alpha \le \frac{h(n)}{n} \le (\alpha + \varepsilon)(1 + \varepsilon) + \varepsilon,$$

Q.E.D.

Claim 1.5.2.
Let S be a stationary combinatorial source over an alphabet A, $h(l) = H(S^1 \vee \ldots \vee S^l)$, if $S \subseteq A^\infty$, $h(l)$ be the logarithm of the quantity of l -length words in S, $S \subseteq A^*$. Then there exists the limit per letter entropy

$$H(S) = \lim_{l \to \infty} \frac{h(l)}{l}, \qquad H(S) \le \frac{h(l)}{l}.$$

The Claim follows Claim 1.5.1.
In accordance with Section 1.1, the source is called nontrivial, if its limit per letter entropy is not zero.

Claim 1.5.3.
Let S be a stationary combinatorial source over A. Then its limit per letter entropy $H(S)$ does not exceed $\log_2 |A|$. It equals $\log_2 |A|$ iff S includes all words ($S = A^*$).

Proof.
The quantity of l-length words is not greater than $|A|^l$, and the inequality

$$H(S) \le \log |A|$$

is obvious. Suppose there is a word w, $|w| = a$, $w \notin S$. Then the quantity of a-length words of S is not greater than $|A|^a - 1$:

$$h(a) \le \log\left(|A|^a - 1\right). \tag{1.5.8}$$

From (1.5.8) and the inequality $h(a + b) \le h(a) + h(b)$ we get for any $q > 0$:

$$\frac{h(qa)}{qa} \le \frac{\log\left(|A|^a - 1\right)}{a} < \log |A|.$$

Thus,

$$H(S) = \lim_{q \to \infty} \frac{h(qa)}{q(a)} < \log |A|,$$

Q.E.D.

There is a special kind of stationary combinatorial sources. They are called Jablonskii invariant classes and consist of boolean functions.

A boolean function f is a map of the set E^n of all n-length binary words to the set $\{0, 1\}$, $n \ge 1$. A function f is represented by the 2^n-length word of values, which f takes at all x, $x \in E^n$. The words x are ordered lexicographically: the words, whose first letter is 0, go first, etc. A variable x_i, $1 \le i \le n$, is called fictitious for a function f, if

$$f(x_1, \ldots, x_{i-1}, 1, x_{i+1}, \ldots, x_n) = f(x_1, \ldots, x_{i-1}, 0, x_{i+1}, \ldots, x_n).$$

Two functions are equivalent, if one of them can be obtained from the other through either dropping fictitious variables or permuting variables.

A set S of boolean functions is called a Jablonskii invariant class, if

a) it follows from $f \in S$ that all functions equivalent to f belong to S

b) it follows from $f \in S$ that all functions obtained by substituting some constants into f belong to S.

Particularly, if $f(x_1, \ldots, x_n) \in S$, then both $f(1, x_2, \ldots, x_n)$ and $f(0, x_2, \ldots, x_n)$ are elements of S. The 2^{n-1}-length word of values of $f(0, x_2, \ldots, x_n)$ is the left half of that word of $f(x_1, \ldots, x_n)$. Functions being identified with their words of values, we can say, that if a word belongs to a class S, then its left and right halves belong to S as well. If $h(2^n)$ is the quantity of 2^n-length words (functions of n variables) in a class S, then

$$h(2^n) \le 2h(2^{n-1}).$$

Claim 1.5.1 yields that there exists

$$\lim_{n \to \infty} \frac{\log h(2^n)}{2^n} = \sigma.$$

Claim 1.5.3 yields that $\sigma = 1$ iff S includes all boolean functions

S. Jablonskii proved that for any σ, $0 \le \sigma < 1$, there is continuum different invariant classes, whose limit per letter entropy equals σ.

Concluding this Section we find limit per letter entropy for some invariant classes of common use.

1. Class S of linear function. A function $f \in S$, if

$$f = C_0 + C_1 x_1 + \ldots + C_n x_n \pmod 2, \quad C_i = 0, 1, \ i = 1, \ldots, n.$$

A function f is specified by the constants C_i. Hence, the number $h(2^n)$ of the functions of n variables is not more than 2^{n+1},

$$\sigma = \lim_{n \to \infty} \frac{\log h(2^n)}{2^n} = 0.$$

2. Class S of symmetric functions, which are invariant under any permutation of non fictitious variables. To specify such a function of n variables, one shall specify its values on a word with i ones, for any i, $0 \le i \le n$. Hence, the number of such functions of n variables is not more than 2^{n+1}, and limit per letter entropy is zero.

3. Class S of monotone functions. A word $x = (x_1, \ldots, x_n)$ is not less than a word $y = (y_1, \ldots, y_n)$, if $x_i \le y_i$, $i = 1, \ldots, n$. A function f is monotone, if $f(x) \le f(y)$ for any x, y, $x \leq y$. The number $h(2^n)$ of monotone functions of n variables is known to be not more than $n^{C_n^{n/2}}$. Hence,

$$\sigma = \lim \frac{\log h(2^n)}{2^n} \leq C \cdot \frac{2^n \log n}{\sqrt{n} 2^n} = 0.$$

So, the limit entropy of all three those classes is zero.

NOTES

The entropy of a combinatorial source was introduced by R. V. L. Hartley (1928). It took 20 years to develop a corresponding concept for stochastic sources. It was done by C. E. Shannon. His pioneering paper (1948) is the foundation of information theory.

Epsilon-entropy is a generalisation of the Hartley entropy on metric spaces. Section 1.2 is an exposition of results of Kolmogorov and Tichomirov (1959).

Markov sources are named after their inventor A. A. Markov. Studying the text of the poem "Eugene Onegin" he observed that the probability to meet a consonant or a vowel depends on whether the preceding letter was a consonant or a vowel. That observation gave rise to the theory of Markov chains. The theory is still flourishing. One can say that A. A. Markov used first order Markov source with two states as a model of the poem. The first state is "vowel"; the second one is "consonant". Either a consonant or a vowel is generated in any state .

Markov's idea has found its continuation in Shannon (1948). In that paper English texts were considered as Markov sources of different order. Depending on the order the entropy equals

H_{comb}	H_0	H_1	H_2	...	H_4	H_7
$4, 76$	$4, 03$	$3, 32$	$3, 10$	...	$2, 1$	$1, 9$

By H_{comb} we mean the entropy of the combinatorial source, which generates 27 English letters, $H_0 = \log_2 27$. By H_0 we mean the entropy of the Bernoulli source which generates the English letters with their stationary probabilities, and so on.

Markov sources are employed to describe documents to be sent via fax, TV-pictures etc.

Jablonskii (1959) introduced invariant classes of boolean functions.

CHAPTER 2

Source Coding

Main types of compressing maps (encodings) are introduced. The Kraft inequality is a criterion of decipherability. The average number of coding letters per an input letter is the encoding cost. The difference between the encoding cost and the source entropy is the redundancy, which is nonnegative for decipherable codes. So, the entropy is the cost of an ideal encoding. There are different methods to bring the cost closer to the entropy. For none of them the redundancy of a Markov source can be less than $\frac{\text{const}}{n}$, n being the coding delay. The only exception is the Bernoulli source with concerted probabilities.

There is an interesting relation between threshold boolean functions and decipherable codes.

Allow a minor violation of decipherability: let several words obtain one and the same code. We get a method of loading dictionaries to computer memory, which is called hashing.

A word may be given codes of different length depending on which combinatorial source it is generated by.

There is a universal or Kolmogorov encoding which is just as good as any other (to within an additive constant). All conventional codes are majorants of Kolmogorov's.

If a stationary source generates the most complicated words, then it generates each word.

2.1 Types of codes

To compress data one changes words generated by a source for their codes. The code should be shorter in a sense than the word. The coding should be decipherable: one can recover the word by its code.

The encoder is a finite automaton. To produce the code of a word w it reads it letter by letter going from one state to another. In any state it prints a word of the output alphabet. That word may be empty. An encoder is said to be information lossless, if through, first, its initial state, second, the code of a word w, third, the state, which the encoder gets in after reading w, one can find w. It is said to be information lossless of finite order if there is an integer m such that through the initial state and the code of word w, $|w| = m$, one can find the first letter of w. The encoders we are going to deal with are information lossless of finite order. That class is broad enough for both theoretical and practical purposes.

If all words printed by an automaton in all its states are of the same length, we get a fixed rate encoding. If not all of them are of the same length, we get a variable rate encoding. Take an information lossless automaton f with σ states, $\sigma \geq 1$, and a word w in an alphabet A. Suppose n is a divisor of $|w|$, $n \geq 1$. Being fed with an n-length word x, $x \in A^n$, in a state σ_j, $j = 1, \ldots, \sigma$, the automation f produces a binary word w_{xj}. If the encoding is fixed rate, then all words w_{xj} have equal length:

$$|w_{xj}| = l, \quad l \geq 1.$$

Denote by l_x the minimal length of all codes, which a letter x has got:

$$l_x = \min_{1 \leq j \leq \sigma} |w_{xj}|.$$

Then we get the following lower bound for the per letter codelength $C(f, w)$ of the word w :

$$C(f, w) \geq \frac{1}{n} \sum_{x \in A^n} \frac{r_x(w)}{|w|} l_x, \tag{2.1.1}$$

where $r_x(w)$ is the number of nonoverlapping occurences of a word $x \in A^n$ in w.

Per letter fixed rate codelength of w equals

$$C(f, w) = \frac{l}{n}. \tag{2.1.2}$$

A special attention will be focused on a subclass of automaton codes, which are defined through oriented trees. To introduce that subclass we give first some definitions, which will be useful for other section of the book as well.

A word x is a prefix of a word y, if $y = xz$, where z is a word, xz is the concatenation of the words x and z. A set of words (code) is called prefix, if none of its words is a prefix of any other of its words. E.g., the set $\{0, 10, 11\}$ is prefix, the set $\{0, 10, 100\}$ is not.

A prefix code can be identified with a binary tree. Any tree consists of internal and external nodes and edges. One of internal nodes is called the root. External nodes are called leaves. Any internal node x is connected by edges with two nodes, which are called its either left or right son. Those nodes are brothers with respect to each other; the node x is their father. Mark the leftgoing edge with 0, the rightgoing

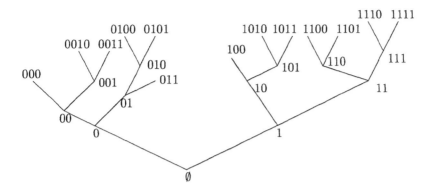

Fig. 2.1.1. A binary tree.

one with 1. Then each node will be provided with a binary word describing how to get to this node from the root. The word the root is provided with is empty (Fig. 2.1.1). We will not make distinction between a node and the word it is provided with.

If Δ is a tree, x is a node of Δ, then $\Delta(x)$ is a subtree of Δ, whose root is x. The set of leaves of $\Delta(x)$ is denoted by $L\Delta(x)$. The leaves $L\Delta(\emptyset)$ of Δ make a prefix code. On the other hand, the words of any binary prefix code are the leaves of a tree.

The root of a tree makes its first storey of level; its sons make the second storey etc. The number of storeys of a tree is called its height. A tree is uniform, if the quantities of leaves of any two subtrees, whose roots belong to the same storey, differ not more than by 1. The tree on Fig. 2.1.1 is uniform. The inequality

$$|L\Delta(x)| \leq |L\Delta(\emptyset)|\, 2^{-|x|} + 1 \tag{2.1.3}$$

holds for a uniform tree, where $|L\Delta(x)|$ is the number of leaves of the tree $\Delta(x)$, $|x|$ is the length of the word x (the number of the storey, which x belongs to).

There is the lexicographic order on the set of all binary words. The words, whose first letter is 0, precede the words, whose letter is 1, and so on. The nodes of a tree are ordered the same way. The leaves are ordered linearly, that is, they make a chain. The nodes belonging to a storey make a chain too.

A tree Δ has got a precedence function Prec: Prec x is the quantity of leaves of Δ preceding its node x. For the tree Δ depicted on Fig. 2.1.1, we have Prec $0 =$ Prec $(0) =$ Prec $(00) =$ Prec $(000) = 0$, Prec $(1) = 6$, Prec $11 = 9$.

Let $x = x_1 \ldots x_{|x|}$ be a node of a tree Δ. As it is easily seen,

$$\text{Prec } x = x_1 \left|L\Delta(0)\right| + x_2 \left|L\Delta(x_1 0)\right| + \ldots + x_{|x|} \left|L\Delta(x_1 \ldots x_{|x|-1} 0)\right| \tag{2.1.4}$$

As it follows from (2.1.4), for any two nodes x, y, $x < y$

$$\text{Prec}\,(y) \geq \text{Prec}\,(x) + |L\Delta(x)| \tag{2.1.5}$$

For any node x

$$\text{Prec}\,(x) + |L\Delta(x)| \leq |L\Delta(\emptyset)| \tag{2.1.6}$$

A set $P = \{\Delta_1, \ldots, \Delta_1 P_1\}$ of disjoint subtrees of a tree Δ is called a partition of Δ, if every leaf of Δ is a leaf of a tree Δ_i, $i = 1, \ldots |P|$. The subtrees $\{\Delta(000), \Delta(001), \Delta(01), \Delta(1)\}$ make a partition or the tree on Fig. 2.1.1. The roots of the subtrees of P are the leaves of a subtree Δ' of Δ. The subtree Δ' is called the basis of P. It is clear that

$$|P| = |L\Delta'| \tag{2.1.7}$$

$$\sum_{i=1}^{|P|} |L\Delta_i| = |L\Delta| \tag{2.1.8}$$

For any binary tree Δ the number of its leaves and the number of its nodes $|\Delta(\emptyset)|$ are related by (2.1.9):

$$|L\Delta(0)| = \frac{1}{2} \left(|\Delta(\emptyset)| + 1 \right) \tag{2.1.9}$$

Now we go back to the automaton codes which are defined through trees. Let Δ be a tree, whose any node has got k sons, $k \geq 2$. Provide a leaf x with a binary code $f(x)$. The words $\{f(x)\}$ make a prefix set. Take a word w over a k-letter alphabet. There is a shortest prefix w' of w which describes a path from the root of Δ to a a leaf x'. Get the code $f(x')$, omit w'. We get w'', $w = w'w''$. Move from the root according to w''. As soon as a leaf is reached, take its code and concatenate it with $f(x')$. Omit the corresponding prefix of w'' and go on. If such a travel is terminated at a leaf, we will get the code of w. If it is terminated at an internal node, we will get the code of a prefix of w. Add to w several extra letters. Then the new word will get a code too. The encoding is defined by the tree Δ and the set $\{f(x)\}$. It may be an information lossless automaton of a finite order. Let m be the height of the tree Δ, w be a word, $|w| \geq m$. Then one can find the first letter of w by its code. There are four types of encodings:

- BB (block to block)

- BV (block to variable length)

- VB (variable length to block)

- VV (variable length to variable length).

The first letter of the type of a code is either B or V, depending on whether or not all distances from the root of the tree Δ to its leaves are equal. The second letter is either B or V, depending on whether or not all lengths of the codewords $\{f(x)\}$ are equal.

If all distances of leaves of Δ are equal to a number n, $n \geq 1$, then we get a BV or BB encoding. To encode a word w one has to subdivide it into nonoverlapping n-length subwords and to give a subword x a code $\{f(x)\}$, $|f(x)| = l_x$. Inequality (2.1.1) turns to an equality:

$$C(f, w) = \frac{1}{n} \sum_{x \in A^n} \frac{r_x(w)}{|w|} |f(x)|, \qquad (2.1.10)$$

where $C(f, w)$ is the length of the code of a word w. For BB-encoding equality (2.1.2) remains valid.

When encoding a stationary source S, a word x has got a probability $p(x)$.) The length of the code of x is $|f(x)|$, the average per letter codelength or the encoding cost is

$$C(f, S) = \frac{1}{n} \sum_{x, x \in A^n} |f(x)| p(x) \qquad (2.1.11)$$

If not all those distances are equal, then the encoding is not block. The encoding procedure has already been described: travel over Δ according to a word w, reach a leaf, take its code, start again from the root. Let $f(w)$ be the code of the longest prefix of w, which corresponds to a leaf. Then it is natural to define the cost of the encoding f on a source S as

$$C(f, S) = \lim_{n \to \infty} \frac{1}{n} \sum_{w, |w|=n} p(w) |f(w)|. \qquad (2.1.12)$$

We find a simple explicit formula for the cost of encoding of a Bernoulli source. The delay, or the average height of a tree Δ, equals

$$d = \sum_x p(x) |x|, \qquad (2.1.13)$$

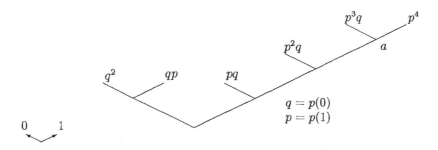

Fig. 2.1.2. Probabilities of leaves of a binary tree.

where the sum is taken over all leaves of the tree Δ. So, for the tree of Fig. 2.1.2, its delay or average height equals

$$d = 2q^2 + 4pq + 3p^2q + 4p^3q + 4p^4$$

where $p = p(0)$ is the probability of one, q is the probability of zero.

Lemma 2.1.1. (average height of a tree)

Let Δ be a tree, $S = \{A_1, \ldots, A_k\}$ be a Bernoulli source, $|x|$ be the distance of a node x from the root, $p(x)$ be the probability for a word x to be generated by S, $r_j(x)$ be the number of occurences of the letter A_j in x, $j = 1, \ldots, k$. Then

$$\sum_{x, x \text{ is a leaf of } \Delta} p(x)|x| = \sum_{y, y \text{ is a node of } \Delta} p(y)$$

$$- \sum_{x \text{ is a leaf of } \Delta} p(x) \log p(x) = H(S)d$$

$$\sum_{x \text{ is a leaf of } \Delta} p(x)r_j(x) = dp(A_j), \quad j = 1, \ldots k.$$

Proof.

Take a tree Δ. Suppose that the sons of a node a are leaves. Remove them. A tree Δ' is obtained (see Fig. 2.1.2). Proceed by induction.

Suppose that Lemma 2.1.1 holds for Δ'. Prove it for Δ. We will do it for the first equality. The other two are dealt with the same way. S being a Bernoulli source, we have

$$\sum_{x \text{ is a son of } a} p(x) = p(a)p(A_1) + \ldots + p(a)p(A_k) = p(a). \tag{2.1.14}$$

The equality (2.1.14) yields

$$\sum_{y, y \text{ is a node of } \Delta} p(y) = \sum_{y, y \text{ is a node of } \Delta'} p(y) + \sum_{x, x \text{ is a son of } a} p(x) = \sum_{y, y \text{ is a node of } \Delta'} p(y) + p(a) \tag{2.1.15}$$

Use in (2.1.15) the inductive assumption

$$\sum_{y, y \text{ is a node of } \Delta'} p(y) = \sum_{y, y \text{ is a leaf of } \Delta'} p(y)|y|$$

It gives

$$\sum_{y, y \text{ is a node of } \Delta} p(y) = \sum_{y \text{ is a leaf of } \Delta', \, y \neq a} p(y)|y| + p(a)|a| + p(a) \tag{2.1.16}$$

Relations (2.1.13) and (2.1.15) yield the first statement of the Lemma (by induction), Q.E.D.

Claim 2.1.1. (the cost of VV-code)
The cost $C(f, S)$ of a code f on a Bernoulli source S is

$$C(f, S) = \frac{1}{d} \sum_{x, x \text{ is a leaf of } \Delta} |f(x)| \, p(y)$$

Proof.
Given a tree Δ, the coding procedure is described by a Markov chain. The chain goes from a node x either to the node xA_j, if x is not a leaf, $j = 1, \ldots, k$ or to the node A_j, if x is a leaf. The probability of such an event is $p(A_j)$. Denote by $p^s(x)$ the stationary probability of the state (node)x. Those probabilities meet the following set of equations:

$$p^s(xA_j) = p(A_j)p^s(x) \tag{2.1.17}$$

$$\sum_{x, x \text{ is a node of } \Delta} p^s(x) = 1 \tag{2.1.18}$$

Let

$$p^s(x) = \frac{1}{d} p(x) \tag{2.1.19}$$

d being defined by (2.1.13). Then equalities (2.1.17) are met, since S is a Bernoulli source. Equality (2.1.18) is met by the first statement of Lemma 2.1.1. If the chain works for some time, then it spends an arbitrarily close to p^s fraction of that time in the state x. So, for the cost of the coding (2.1.12) we get

$$C(f, S) = \sum_{x, x \text{ is a leaf of } \Delta} p^s(x) \, |f(x)|$$

From that and (2.1.19) we come to the conclusion of the claim. Q.E.D.

2.2 Kraft Inequality and Levenstein Code

One of the main tools of the source encoding theory is the Kraft inequality.

Claim 2.2.1. (Kraft inequality)
Let $l_1, \ldots, l_k$ be a set of natural numbers, $k \geq 1$. There exists a prefix code, whose wordlengths are $l_1, \ldots, l_k$ iff the following Kraft inequality holds:

$$\sum_{i=1}^{k} 2^{-l_i} \leq 1$$

Proof.
Sufficiency.
Let the Kraft inequality hold, s_i be the quantity of numbers $l_1, \ldots, l_k$, which equal i, $l = \max l_i$, $i = 1, \ldots, k$. The inequality takes form

$$\sum_{i=1}^{l} S_i 2^{-i} \leq 1 \qquad (2.2.1)$$

Let E^i be the set of all binary i-length words, $i = 1, \ldots, l$, $|E^i| = 2^i$. We develop a prefix set S, for which (2.2.1) is met. Include into S any S_1 words of length 1. It is possible, because $S_1 \leq 2^1$. There are $S_1 2^{i-1}$ words of length i, for which those 1-length words are prefixes. Exclude those i-length words from each set E^j, $i = 1, \ldots, l$. There are $2^2 - S_1 2^1$ words of length 2 now. Include any S_2 of them into S. It is possible by (2.2.1). Exclude all words, whose prefixes are those S_2 words, etc. At the l-th step we will get a set S, whose wordlengths are just $l_1, \ldots, l_k$.
Necessity.
Let a prefix set S contains words, whose lengths are $l_1, \ldots, l_k$. Any l_i-length word is a prefix for 2^{l-l_i} words of length l. From the prefix condition we get

$$\sum_{i=1}^{k} 2^{l-l_i} \leq 2^l,$$

which is equivalent to the Kraft inequality, Q.E.D.

Suppose S is a set of words (code). If any words of S can be recovered from their concatenation, then S is called decipherable. A prefix code is decipherable. The code $\{0, 01\}$ is not prefix, but decipherable. Each decipherable code meets the Kraft inequality. It was proved by Macmillan. The following proof was proposed by Karush (1961).

Claim 2.2.2. (decipherability implies Kraft inequality)
If $l_1, \ldots, l_k$ are the wordlengths of a binary decipherable code, $k \geq 1$ then

$$\sum_{i=1}^{k} 2^{-l_i} \leq 1.$$

Proof.
Let a_i be a word of a binary decipherable code, $|a_i| = l_i$, $i = 1, \ldots, k$. Let A be the alphabet, whose letters are $a_1, \ldots, a_k$. Take a word over A and change its letters for the corresponding binary words. We get a binary word. The quantity of those n-letter words over A, which became binary t-letter words after such a change, is denoted by $S(n, t)$, $n \geq 1$, $t \geq 1$. If a word has n letters in A, then it will have not more than nl binary letters after such a change, $l = \max_{1 \leq i \leq k} l_i$. The given code is

decipherable. Thus, t-length binary words generated by different words over A, are different. Consequently,

$$S(n,t) \leq 2^t$$

Let

$$x = \sum_{i=1}^{k} 2^{-l_i}, \quad n \geq 1. \tag{2.2.2}$$

We get

$$x^n = \left(\sum_{i=1}^{k} 2^{-l_i} \right)^n = \sum_{1 \leq i_1 \leq \ldots \leq i_n \leq k} 2^{-(l_{i_1} + \ldots + l_{i_n})} \tag{2.2.3}$$

The sum $l_{i_1} + \ldots + l_{i_n}$ is the binary length of the word $a_{i_1} a_{i_2} \ldots a_{i_n}$. That sum will take the value t exactly $S(n,t)$ times. We rewrite (2.2.3) as

$$x^n = \sum_{t \leq nl} S(n,t) 2^{-t} \tag{2.2.4}$$

From (2.2.2) and (2.2.4) we get

$$x^n \leq nl \tag{2.2.5}$$

The inequality (2.2.5) holds, if n goes to infinity. Thus, x can not exceed 1, and $x \leq 1$, Q.E.D.

Go from prefix codes to a more general case of automaton ones. The Kraft inequality ought to be slightly corrected. We take an information lossless automaton with σ states, $\sigma \geq 1$, and a t-letter alphabet A, $t \geq 1$. Being fed with a letter A_i in a state σ_j the automaton produces a binary word w_{ij}, $i = 1, \ldots t$, $j = 1, \ldots \sigma$. For fixed rate encoding we have

$$|w_{ij}| = l.$$

Let l_i be the minimal lengths of all codes of a letter A_i :

$$l_i = \min_{1 \leq j \leq \sigma} |w_{ij}|$$

Claim 2.2.3. (Lempel - Ziv - Kraft inequality)
Let A be an alphabet, $|A| = t$, $t \geq 1$, $\{l_i\}$ be the set of minimal length of a binary information lossless automaton code. The encoding automaton has σ, $\sigma \geq 1$ states. Then

$$K = \sum_{i=1}^{t} 2^{-l_i} \leq \sigma^2 \left(1 + \log \frac{\sigma^2 + t}{\sigma^2} \right)$$

For a fixed rate encoding $l_x = l$, $l \geq \log \frac{t}{\sigma^2}$
Proof.
The symbol s_j stands for the quantity of letters for which $l_i = j$, $j = 0, 1, \ldots$. Then

$$K = \sum 2^{-l_i} = \sum_j s_j 2^{-j} \tag{2.2.6}$$

and

$$\sum_{j=0} s_j = t. \tag{2.2.7}$$

The encoding automaton is information lossless. Hence, if one knows, first, the state, where the automaton sees a letter, second, the state, where it goes, third, the code of this letter, then one can find the letter. Consequently,

$$s_j \leq \sigma^2 t^2 \tag{2.2.8}$$

If the encoding is fixed rate, then

$$l_1 = \ldots = l, \quad s_l = t$$

and (2.2.8) is equivalent to the second statement of the Claim.

To obtain an upper bound on K we may overestimate initial numbers s_j at the expence of the following ones, provided the sum of all of them remains equal to t. Thus, we can rewrite (2.2.6) using (2.2.7):

$$K \leq \sum_{j=0}^{m} \left(\sigma^2 \cdot 2^j \right) \cdot 2^{-j} = \sigma^2 (m+1). \tag{2.2.9}$$

In (2.2.8) m is the integer satisfying the inequality

$$\sum_{j=0}^{m-1} \sigma^2 \cdot 2^j < t \leq \sum_{j=0}^{m} \sigma^2 \cdot 2^j.$$

Furthermore

$$2^m = 1 + \sum_{j=0}^{m-1} 2^j \leq \frac{t}{\sigma^2} + 1$$

which together with (2.2.9) yields the statement of the Claim. Q.E.D.

The Kraft inequality works for both finite and infinite sources. Next we describe the Levenstein code – a decipherable representation of integers. The common binary representation of integers is neither prefix nor decipherable. Say, given a word 111, we can not tell, whether it is the concatenation 1 11 = 13, or 11 1 = 31. Levenstein (1968) gave to the integers a bit more lengthy, but prefix representation. The overhead of such a representation is minimal.

Let Bin x be the usual binary representation of a number x, Bin $'x$ be Bin x after dropping its first digit:

$$\text{Bin } 3 = 11, \quad \text{Bin } '3 = 1, \quad \text{Bin } '2 = 0, \quad \text{Bin } '1 = \emptyset.$$

Evidently,

$$|\text{Bin } 'x| = \lfloor \log x \rfloor, \quad x \geq 2.$$

Define

$$n^{(i)} = 2^{n^{(i-1)}}, \quad i = 1, 2, \ldots, \quad n^{(0)} = 1.$$

Then

$$n^{(1)} = 1, \quad n^{(2)} = 2, \quad n^{(3)} = 4, \quad n^{(4)} = 16, \quad n^{(5)} = 2^{16}, \quad \ldots .$$

Define

$$\log^* x = i, \quad \text{if } n^{(i)} \leq x \leq n^{(i+1)}.$$

The function $\log^* x$ grows extremely slow. For instance,

$$\log^* 100 = 4, \quad \log^* \left(2^{10^4}\right) = 5.$$

The Levenstain code Lev x of a number x is defined as follows. Take the word Bin $'x$. Write to the left of it the word Bin $'\lfloor \log x \rfloor$, to the left of it – Bin $'\lfloor \log \log x \rfloor$, etc, all in all $\log^* x - 1$ times. Then write 0 and $\log^* x$ ones. E.g., for

$$x = 37 \quad \lfloor \log x \rfloor = 5, \quad \lfloor \log^{(2)} x \rfloor = \lfloor \log \log x \rfloor = 2, \quad \lfloor \log^{(3)} x \rfloor = 1, \quad \log^* x = 4,$$

$$\text{Bin } '1 = \emptyset, \quad \text{Bin } '2 = 0, \quad \text{Bin } '5 = 01, \quad \text{Bin } '37 = 00101, \quad \text{Lev } 37 = 111000100101;$$

$$\text{Lev } 1 = 10, \quad \text{Lev } 2 = 1100, \quad \text{Lev } 4 = 1110000,$$

$$\text{Lev } 8 = 11101000, \quad \text{Lev } 16 = 111100000000.$$

It is easy to recover x from Lev x. We explain it through an example. Suppose we have a word, whose beginning is 111100100010111101001. We can see that $\log^* x = 4$. Take the first digit after the first zero, write 1 to the left. We get $10 = 2$. Take two more digits, write 1: 110=6. Take six more digits, write 1 to the left: 11001011 = 75. The procedure was repeated $\log^* x - 1 = 3$ times, so $x = 75$. The number x can be separated from any concatenation. The Levenstein code is decipherable. We get for its length:

$$|\text{Lev } x| \sim \log x \tag{2.2.10}$$

$$|\text{Lev } x| = \log x + \log \log x (1 + O(1)) \tag{2.2.11}$$

The Levenstein code meets the Kraft inequality.

2.3 Encoding of Combinatorial Sources

Let S be a finite combinatorial source, $H(S) = \log |S|$ be its Hartley entropy, f be an injective block code. Any word $A \in S$ is given an m-length code $f(A)$, $|f(A)| = m$, $f(A_i) \neq f(A_j)$, $i \neq j$. The redundancy of f on S is

$$R(f, S) = m - H(S).$$

Dividing the redundancy $R(f, S)$ by either m or $H(S)$ we get two types of relative redundancies:

$$\rho(f, S) = \frac{R(f, S)}{m},$$

$$\rho'(f, S) = \frac{R(f, S)}{H(S)}.$$

Evidently,

$$\rho'(f, S) = \frac{R(f, S)}{1 - \rho(f, S)}.$$

Claim 2.3.1.

If S is a finite combinatorial source, f is a block injective code, then the redundancy of f on S is nonnegative. There is a block code f, whose redundancy is not greater than 1.

Proof.
The quantity of m-length words is 2^m. The map f being injective,

$$2^m \geq |S|.$$

It gives the first statement of the Claim. Taking

$$m = \lceil |S| \rceil,$$

we get a map f,

$$R(f, S) \leq 1.$$

Q.E.D.
   Block encoding of a source is a model of information retrieval. The words of S are called keys. A key $x \in S$ is put to a computer cell, whose address is $f(x)$. The cells, whose addresses are

$$\{0, 1, \ldots, 2^m - 1\}, \quad |f(x)| = m,$$

constitute a table. The number

$$\alpha(f, S) = \frac{|S|}{2^m}$$

is called the loading factor of the table. Obviously,

$$\alpha(f, S) = 2^{-R(f,S)}.$$

The coding map f is called key-address transformation.

There are two types of key-address transformation. For the first type $f(x) = 0$, if $x \in S$. In that case one can tell through $f(x)$ whether or not x belongs to S. Such transformations are called strong. For the second type one can not tell through $f(x)$ whether or not $x \in S$. Such transformations are called weak.

If an injective map is not block, then there can be codewords, whose length is less than the entropy of S. However, such codewords make an exception in a sense.

Claim 2.3.2.
If f is an injective map of S to the set of binary words, then there is $x \in S$ for which

$$|f(x)| \geq \lfloor \log |S| \rfloor.$$

For any t such that
$$\lfloor \log |S| \rfloor \geq t \geq 1,$$

the words $x \in S$, for which
$$|f(x)| \leq \lfloor \log |S| \rfloor - t,$$

constitute not more than $2^{-(t-1)}$ th-fraction of the set S.

Proof.
The number of binary words, whose length is not more than $\lfloor \log |S| \rfloor - t$, is

$$2^1 + \ldots + 2^{\lfloor \log |S| \rfloor - t} \leq 2^{1-t} |S|.$$

If yields the second statement of the Claim. We obtain the first statement letting $t = 1$. Q.E.D.

An example of the nonblock source encoding is computation of functions. Programs of functions play role of their codes. We will use random access machines. Program length will be measured in bits. Running time is understood as the number of bitoperations.

Let g be a map of a finite set of binary words dom g to a finite set of binary words Img . Load into the memory of a computer a set P of binary words. Load into a special cell a word $x \in$ dom g and start the machine. If, after a while, the machine stops and prints $g(x)$, then P is called a program of g. The minimal bitlength of programs of g is program complexity $\Pr(g)$ of g. The maximal over $x \in$ dom g number of bitoperations is time of calculation of g given a program P. It is denoted by $T(g)$. The symbol $S(g)$ stands for the additional memory necessary to find g given P.

Claim 2.3.3.
For any boolean function of n variables g there is a calculating program P whose bitlength $\Pr(f)$ is not more then $2^n + c$, $c = \mathrm{const}$, running time $T(g)$ is $n + 1$. The program P needs not any additional memory. There are functions, whose program complexity is not less than 2^n.

Proof.
Take 2^n one-bit cells of a computer. The cell with the number val x, $x \in E^n$, is loaded with $g(x)$. Those one-bit cells are called the informational part of a program P. There is an n-length cell to be loaded with an input word x, $x \in E^n$. There is also a computer instruction: find the cell with the number val x, x being the input word, and print the content of that cell. This instruction may be called operational part of P. The program length is $2^n + c$, c being the program length of the operational part. As it follows from Claim 2.3.2., there are functions with programlength not less than $2^n (|S| = 2^{2^n}$, f maps a function to its program). To compute $f(x)$ we read x (n operation) and extract $f(x)$ from the memory (one operation). Q.E.D.

Such a method of calculation of boolean functions is called tabular. Those cells make a table of a function. The method is quite simple. Still, there are no simpler ways to calculate some boolean functions. The calculation of monotone functions is a closely related topic. There is a simple two-level method to compute them.

Claim 2.3.4.   (two-level computation of monotone functions by V. Potapov)
Let g be a monotone integer-valued function, specified on a finite segment $[1, 2, \ldots, M]$ $M \geq 1$. Suppose that its jumps do not exceed a number p :

$$g(i) - g(i-1) \leq p, \quad i = 1, \ldots, p-1.$$

There is a table, whose size is

$$\log g(1) + M(1 + \log p + \log \log Mp).$$

It takes two indexed memory reads and two additions to find $g(x)$, $x = 1, \ldots, M$.
Proof.
Make a first level table, whose i-th word is

$$g(i \log Mp) - g(1), \quad i = 1, 2, \ldots, \frac{M}{\log Mp}.$$

The length of its words is $\log Mp$, since

$$g(x) - g(1) \leq Mp, \quad x = 1, \ldots, M.$$

The size of the table is M bits.
Make a second level table whose i-th word is

$$g(i) - g\left(\left[\frac{i}{\log Mp}\right] \log Mp\right), \quad i = 1, \ldots, M.$$

The length of its words is $\log p \log Mp$, since

$$g(i) - g\left(\left\lceil \frac{i}{\log Mp} \right\rceil \log Mp\right) \leq p \log Mp.$$

The size of the table is

$$M(\log p + \log \log Mp).$$

Write $g(1)$, which takes $\log g(1)$ bits. The combined bitlength of the both tables and $g(1)$ is as required. We find $f(x)$ in an obvious way through those tables. Q.E.D.

2.4 Hashing

When discussing key-address transformations, we understood them to be injective maps. Different words are given different addresses. Sometimes it is beneficial to employ nearly injective maps. Two different words may be provided with one and the same address. We say that a collision has occurred in this case. Let S be a source (dictionary), B (table)

$$B = \{0, \ldots |B| - 1\}$$

be a set, f be a map $S \to B$. The set of words, which are given an address k, $k \in B$, is called a cluster. It is

$$\left\{f^{-1}(k) \cap S\right\},$$

its cardinality is

$$\left|f^{-1}(k) \cap S\right|.$$

A vector $i = i(f, S)$, whose coordinates are the cardinalities of clusters, is called the signature of f on S :

$$i = i(f, S) = (i_0, i_1, \ldots, i_{|B|-1}), \quad i_k = \left|f^{-1}(k) \cap S\right|, \quad k = 0, \ldots, |B| - 1.$$

Obviously,

$$|i(f, S)| = \sum_{k \in B} i_k = |S| \tag{2.4.1}$$

The number

$$I(f, S) = \frac{1}{|S|} \sum_{k \in B} i_k^2 - 1 \tag{2.4.2}$$

is called the colliding index of f on S. If f is an injection, then $I(f, S)$ takes its minimal value-zero. If f is the most noninjective map, i.e., it gives to all words one and the same address, then $I(f, S)$ takes its maximal value - $|S| - 1$. So, the index shows how far is a map from an injection. A function f may be used to load a dictionary into computer memory. In such a capacity f is called a hash-function. An injection is called a perfect hash-function.

The words of a cluster make a list or a chain. Chains are kept separately. To find the first word of the k-th chain, we spend a unit of time, $k = 0, \ldots, |B| - 1$. To find the second word, we spend two units, etc. All in all, we spend

$$1 + \ldots + i_k = \frac{i_k(i_k + 1)}{2}$$

units or time. The average time to find a word of a dictionary S is

$$t(f, S) = \frac{1}{|S|} \sum_{k \in B} \frac{i_k(i_k + 1)}{2}.$$

We see that the average time is linearly related to the index:

$$t(f, S) = \frac{1}{2} I(f, S) + 1 \tag{2.4.3}$$

There is one more indicator of injectivity. It is the traditional chi-square goodness-of-fit test

$$\chi^2 = \sum_{k \in B} \left(i_k - \frac{|S|}{|B|} \right)^2 \frac{|B|}{|S|} \tag{2.4.4}$$

It is the related to the index:

$$\chi^2 = |B| \, (I + 1) - |S| \tag{2.4.5}$$

We will prove two simple lemmas on the index to be applied later.

A map f makes a collision at a word $x \in S$, if there is $x_1 \in S$, $x_1 \neq x$, such that $f(x_1) = f(x)$.

Lemma 2.4.1. (index and collisions)
Let S be a dictionary, f be a map, which makes e collisions, $0 \leq e \leq |S|$, $I(f, S)$ be the index of f on S. Then

$$I(f, S) \geq \frac{e}{|S|}.$$

Proof.
Let A_1, A_2 be the images of, correspondingly, noncolliding and colliding words of S. It means that

$$\left| f^{-1}(k) \right| = 1, \quad k \in A_1,$$

$$\left| f^{-1}(k) \right| \geq 2, \quad k \in A_2. \tag{2.4.6}$$

By the condition of Lemma,

$$\sum_{k \in A_2} \left| f^{-1}(k) \right| = e \tag{2.4.7}$$

We have from (2.4.6) and (2.4.7), that

$$|A_2| \leq \frac{e}{2} \tag{2.4.8}$$

Now we rewrite definition (2.4.2) of the index as

$$I(f, S) = \frac{1}{|S|} \left(|S| - e + \sum_{k \in A_2} \left| f^{-1}(k) \right|^2 \right) - 1 \qquad (2.4.9)$$

Take the Iensen inequality for the function x^2 :

$$x_1^2 + \ldots + x_m^2 \geq (x_1 + \ldots + x_m)^2 \cdot \frac{1}{m} \qquad (2.4.10)$$

Apply (2.4.10) to the sum in (2.4.9). We get

$$I(f, S) \geq I(f, S) = \frac{1}{|S|} \left(|S| - e + \frac{e^2}{|A_2|} \right) - 1 \qquad (2.4.11)$$

The statement of the Lemma follows from (2.4.11). Q.E.D.

If $I(f, S) < \frac{1}{|S|}$, then, by the Lemma, the number of collisions is zero, i.e., f is an injection. We will get a little more precise statement. If a function f is not an injection, then there are at least two elements in one of clusters. To obtain the minimal value of the index, any other cluster should contain not more than one element, as it is easy to verify. So, the minimal index of a noninjective map is

$$\frac{1}{|S|} \left(2^2 + 1(|S| - 2) \right) - 1 = \frac{2}{|S|}.$$

We have got

Lemma 2.4.2. (condition of injectivity)
If the index of a map on a dictionary S is less than $\frac{2}{|S|}$, then the map is an injection.
The next Lemma relates index to the maximal cluster.

Lemma 2.4.3. (maximal size of a cluster)
Let $I(f, S)$ be the index of a map $f : S \to B$ on a dictionary S. Then $\max_{k \in B} |f^{-1} \cap S| \leq \sqrt{I(f, S) |S|} + 1$.
Proof.
Let m be the maximal cardinality of clusters

$$m = \max_{k \in B} \left| f^{-1} \cap S \right|$$

The index $I(f, S)$ will be minimal, if every one of the remaining $|S| - m$ words is given an individual address:

$$I(f, S) \geq \frac{1}{|S|} (m^2 + |S| - m) + 1 \qquad (2.4.12)$$

We get from (2.4.12):

$$|S| \, I(f, S) \geq m(m - 1) \geq (m - 1)^2.$$

Hence, $\sqrt{|S| \, I(f, S)} + 1 \geq m$, Q.E.D.

Given a dictionary S, it is desirable to get an easy to compute hash-function f, whose index $I(f, S)$ is as small as possible. Such a function ought to be either precomputed or taken at random. Common hash-methods are the following one:

1. Squaring.
Several digits of x^2 make the address for a key $x \in S$.

2. Division.
Choose a number k, which is either a prime or does not have any small divisors. Divide $x \in S$ by k. The remainder is the address.

3. Polinomial division.
Choose a polinomial k. Consider $x \in S$ as a polinomial. Divide x by k. The remainder is the address.

4. Digital.
Choose several digits of x.

Those methods are considered to be satisfactory. Still, for every one of them there is a bad dictionary, for which the method gives great indices and many collisions. Our aim is to develop small families of hash functions. For any dictionary there is a function with not very great index in the family. All functions are fast to calculate. The problem will be discussed in Chapter 4.

2.5 Encoding of Individual Words and Stochastic Sources

Let $A = \{A_1, \ldots, A_k\}$ be an alphabet, w be a word over A, f be an automaton encoding. The redundancy of f on w is the difference between the cost of f (see 2.2) and the empirical entropy of w with respect to A^n, $n \geq 1$:

$$R(f, w) = C(f, w) - H_n(w), \qquad (2.5.1)$$

where n is an integer, $n \geq 1$.

Given a stochastic source S, we define the redundancy of f on S the same way:

$$R(f, w) = C(f, S) - H(S) \qquad (2.5.2)$$

If the encoding f is fixed rate, then in the definition of the redundancy we take the combinatorial entropy:

$$R(f, w) = \frac{l}{n} - CH_n(w), \qquad (2.5.3)$$

where l is the codelength of any n-length word, $CH(w)$ is defined in 1.1.

Fixed rate encoding of stochastic sources is out of the scope of this book.

Claim 2.5.1.

Let S be a stochastic source, $S = \{A_1, \ldots, A_k\}$, w be a word, f be an automaton encoding. Then for any $\varepsilon > 0$ and big enough n

$$R(f, w) \geq -\varepsilon$$

and

$$R(f, S) \geq -\varepsilon.$$

If f is a prefix code, then

$$R(f, S) > 0, \quad R(f, w) \geq 0.$$

Proof.

Use (2.1.1.) and the definition of the empirical entropy. We get

$$R(f, w) \geq \frac{1}{n} \sum_{x \in A^n} \frac{nr_x(w)}{|w|} \left(\log 2^{l_x} + \log \frac{r_x(w)n}{|w|} \right). \tag{2.5.4}$$

Here n is a divisor of $|w|$, l_x is the codelength of x, $r_x(w)$ is the number of occurences of x in w,

$$\sum_{x \in A^n} r_x(w) = \frac{|w|}{n}, \quad \sum_{x \in A^n} 2^{-l_x} = k.$$

Apply to (2.5.4) the Iensen inequality for the logarithm. We obtain

$$R(f, w) \geq -\frac{\log k}{n} \tag{2.5.5}$$

Claim 2.2.3 (the Kraft inequality) yields an inequality

$$\log k \leq \log \sigma^2 + \log \left(1 + \frac{k^n}{\sigma^2} \right) \tag{2.5.6}$$

where σ is the number of states of the encoding automaton. If n goes to infinity, then

$$\log k \leq C \log n, \quad C = \text{const} \tag{2.5.7}$$

The inequalities (2.5.5)-(2.5.7) yield

$$R(f, w) \geq -\varepsilon,$$

where $\varepsilon > 0$ is arbitrary, n is big enough. When encoding a source S, we use the probabilities $p(x)$ of words instead of their frequencies $\frac{r_x(w)n}{|w|}$. The result will be the same:

$$R(f, S) \geq -\varepsilon.$$

The Kraft inequality of the Claim 2.2.3 holds for automaton codes. If turns to the inequality of the Claim 2.2.1 for prefix ones. Consequently, zero takes the place of $-\varepsilon$ in the inequalities for prefix codes of Claim 2.5.1. If the encoding is fixed rate, then all codelengths are equal to a number l. There are

$$t = 2^{nCH_n(w)}$$

different nonoverlapping n-length subwords in a word w, see Section 1.1. From the Claim 2.2.3 for fixed rate encoding we obtain, when n goes to infinity:

$$\frac{l}{n} \geq CH_n(w) - \varepsilon,$$

or

$$R(f, w) \geq -\varepsilon$$

(combinatorial case).Q.E.D.

Claim 2.5.1 gives a lower bound for the redundancy. That bound is tight enough. Take the simplest case of fixed rate encoding first. Subdivide a word w into n-length subwords. Let

$$l = \lceil nCH_n w \rceil,$$

where $CH_n w$ is the empirical combinatorial per letter entropy of w. By its definition (Section 1.1), $CH_n w$ is the logarithm of the number of different subwords divided by n. Every subword may be one-to-one encoded by an l-length binary word. The encoding may be done by an automaton. By the inequality

$$l - nCH_n w \leq 1$$

the redundancy does not exceed $\frac{1}{n}$. We have come to

Claim 2.5.2.
Let A be an alphabet, w be a word in A, $n \geq 1$ be a divisor of $|w|$, $CH_n w$ be the empirical combinatorial entropy of w with respect to A^n. Then there is an automaton fixed rate coding f, whose redundancy $R(f, w)$ does not exceed $\frac{1}{n}$.

Nearly optimal variable-length code was developed by C. E. Shannon.

Claim 2.5.3. (the Shannon code)
Let $p(A_i)$ be the probability of a letter A_i, $i = 1, \ldots, k$. There is a prefix code f,

$$A_i \rightarrow f(A_i), \quad |f(A_i)| = \lceil \log p(A_i) \rceil,$$

whose redundancy on S $R(f, S)$ does not exceed 1. If w is a word, then the redundancy of the code f,

$$|f(A_i)| = \left\lceil \frac{\log r_i(w)}{|w|} \right\rceil$$

does not exceed 1 on w :

$$R(f, w) \leq 1.$$

Proof.

We will encode a source S. To encode an individual word w we have just to change the probabilities $p(A_i)$ for the frequencies $\frac{r_i(w)}{|w|}$.

To find the code $f(A_i)$ order the letter $A_1, \ldots, A_k$ in accordance with their probabilities. We can suppose that $p(A_1) \geq \ldots \geq p(A_k)$. Let

$$\sigma_1 = 0, \quad \sigma_2 = p(A_1), \ldots, \sigma_k = p(A_1) + \ldots + p(A_{k-1}),$$

$f(a_i)$ be the first $\lceil -log p(A_i) \rceil$ digits of the binary representation Bin σ_i of σ_i, $i = 1, \ldots, k$. Then

$$p(A_i) \geq 2^{-|f(A_i)|}.$$

One of the first $|f(A_i)|$ digits of Bin $p(A_i)$ is not zero. The numbers $\sigma_{i+1}, \ldots, \sigma_k$ differ from any number $\sigma_1, \ldots, \sigma_i$ by $p(A_i)$ at least. Thus, the words $f(A_i), \ldots, f(A_k)$ differ from $f(A_1), \ldots, f(A_{i-1})$. Consequently, we have got a prefix code, Q.E.D.

The Shannon code is easy to find though it is not optimal. The Huffman code is optimal. It was invented in 1952 and is widely used nowadays.

The Huffman code is based on the following simple Lemma.

Lemma 2.5.1.

Let S be a stochastic source $S = \{A_1, \ldots, A_k\}$, the probabilities are ordered: $p(A_1) \geq \ldots \geq p(A_k)$. There is a code f of minimal cost, whose codelengths do not decrease. The lengths of codes of the two last letters are equal to each other:

$$|f(A_1)| \leq |f(A_2)| \leq \ldots \leq |f(A_{k-1})| = |f(A_k)|.$$

Proof.

Let f be a code for which $C(f, S)$ is minimal. Suppose, on the contrary, that

$$|f(A_i)| > |f(A_j)|, \quad 1 \leq i < j \leq k.$$

Make $f(A_i)$ the code for A_j, $f(A_j)$ the code for A_i. It will lessen the cost - a contradiction. Thus,

$$|f(A_1)| \leq |f(A_2)| \leq \ldots \leq |f(A_k)|.$$

Suppose that

$$|f(A_{k-1})| < |f(A_k)|.$$

Then the brother of $f(A_k)$ is not a codeword and $f(A_k)$ is the only son of its father. Make that father the code of A_k instead of $f(A_k)$. It will decrease the cost - a contradiction. Thus, $f(A_k) = f(A_{k-1})$, Q.E.D.

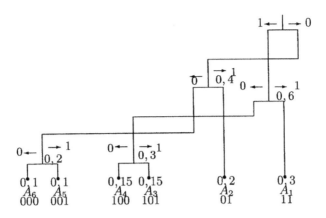

Fig. 2.5.1. Huffman code.

The Lemma gives the following algorithm to construct an optimal code.
a) order the letters of S according to their probabilities
b) make a new alphabet $A'_1, A'_2, \ldots, A'_{k-1}$, and let

$$p(A'_1) = p(A_1), \ldots, p(A'_{k-2}) = p(A_{k-2}), p(A'_{k-1}) = p(A_{k-1}) + p(A_k).$$

Suppose that there is an optimal code f for the new alphabet. Define

$$f(A_1) = f(A'_1), \ldots, f(A_{k-2}) = f(A'_{k-2}),$$

$$f(A_{k-1}) = f(A'_{k-1})0, \quad f(A_k) = f(A'_{k-1})1.$$

Due to the Lemma, this code is optimal.
Example.
Let

$$p(A_1) = 0,3, \quad p(A_2) = 0,2, \quad p(A_3) = p(A_4) = 0,15, \quad p(A_5) = p(A_6) = 0,1.$$

Repeat the procedure a) and b) several times. We get

$$f(A_1) = 11, \quad f(A_2) = 01, \quad f(A_3) = 101,$$

$$f(A_4) = 100, \quad f(A_5) = 001, \quad f(A_6) = 000.$$

The construction of the code is illustrated on Fig 2.5.1.

As D.Huffman said, the letters A_{k-1} and A_k were joined to the letter A'_{k-1} like two brooks were joined to a stream. Then new letters-streams were joined to rivers etc. The codemaking is as the procedure used by a water insect. It goes downstream and makes a mark at joining points to find its way back.

As claim 2.5.1 says, the cost of a decipherable code can not be less then the entropy. Soften the condition of decipherability and take an injective code. It maps different letters to different words, one of which can be a prefix of another. Then the cost may be less than the entropy though not very much.

Claim 2.5.4. (Leung-Ian-Cheong and T.M.Cover (1979), Dunham (1980)).
Let S be a finite stochastic source, $S = \{A_1, \ldots, A_k\}$, $k \geq 1$, f be an injective map, $f(A_i) \neq f(A_j)$, $1 \leq i, j \leq k$. Then

$$C(f, S) \geq H(S) - \log \log k - C,$$

where C is a constant, which does not depend on k.

Proof.
Regard the probabilities of letters as ordered:

$$p(A_1) \geq \ldots \geq p(A_k).$$

The best codes for A_1 and A_2 are 1-length codes 0 and 1. Two-letter word 00, 01, 10, 11 are the best choice for the letters A_3, A_4, A_5, A_6, etc. The length $|f(A_i)|$ of the optimal code for A_i is $\lceil \log \frac{i+2}{2} \rceil$, $i = 1, \ldots, k$. For that code we have

$$H(S) - C(f, S) \leq \sum_{i=1}^{k} p(A_i) \log \frac{2}{p(A_i)(i+2)} \tag{2.5.8}$$

Applying the Iensen inequality to (2.5.8) gives

$$H(S) - C(f, S) \leq \log \sum_{i=1}^{k} \frac{2}{i+2} \tag{2.5.9}$$

To get the Claim we must apply to (2.5.9) the inequality on the partial sum of the harmonic series:

$$\sum_{i=1}^{k} \frac{1}{i} \leq C + \ln k.$$

Q.E.D.

Suppose that the letters of a source S are ordered: $A_1 \leq \ldots \leq A_k$. A code f maps those letters to binary words. A word x is a binary representation of the number val x, Bin val $x = x$. A code f is order-preserving, if from $A_i \leq A_j$ it follows that $f(A_i) \leq f(A_j)$. E.N.Moore and G.N.Gilbert [1959] developed an order-preserving code, which is only slightly worse than Shannon's.

Claim 2.5.5. (Gilbert-Moore order-preserving code)

Let S be a finite stochastic source. There is a prefix order-preserving code, whose redundancy does not exceed 2 (as compared with Shannon's 1).

Proof.

We modify the Shannon code. Let

$$\sigma_1 = \frac{1}{2}p(A_1), \quad \sigma_2 = p(A_1) + \frac{1}{2}p(A_2), \ldots,$$

$$\sigma_i = p(A_1) + \ldots + p(A_{i-1}) + \frac{1}{2}p(A_i), \quad i = 1, \ldots k.$$

Let $f(A_i)$ be the first $\lceil -\log p(A_i)\rceil + 1$ digits of σ_i, $i = 1, \ldots k$. Then

$$R(f, S) = \sum p(A_i)\left((\lceil -\log p(A_i)\rceil + 1) + \log p(A_i)\right) \leq 2. \qquad (2.5.10)$$

As it is clear from (2.5.10),

$$\text{val } f(A_i) < \text{val } f(A_j), \quad i < j,$$

i.e., the code f is order preserving. If $i < j$, then (2.5.10) yields

$$\sigma_j \geq \sigma_i + \frac{1}{2}p(A_i) + \frac{1}{2}p(A_j) \geq \sigma_i + 2^{-|f(A_i)|} + 2^{-|f(A_j)|} \qquad (2.5.11)$$

If $p(A_i) \leq p(A_j)$, then

$$|f(A_i)| = \lceil -\log p(A_i)\rceil + 1 \geq |f(A_j)|.$$

As it follows from (2.5.11), $f(A_j)$ does not coincide with the first $|f(A_j)|$ digits of σ_i. Similarly, if $p(A_i) \geq p(A_j)$, $f(A_i)$ is not a prefix of $f(A_j)$, Q.E.D.

The algorithms of this section may be used to encode an individual word w. We shall use the frequencies $\frac{r_i(w)}{|w|}$ instead of the probabilities $p(A_i)$ for that.

2.6 Equiprobable Letters and Threshold Functions

The simplest source is one with pairwise equal probabilities:

$$S = \{A_1, \ldots, A_k\}, \quad p(A_1) = \ldots = p(A_k) = \frac{1}{k}.$$

The cost of an encoding f is

$$C(f, S) = \frac{1}{k} \sum_{i=1}^{k} |f(i)|.$$

Claim 2.6.1.
The minimal cost of a prefix encoding f of equiprobable letters is

$$C(f, S) = \lfloor \log k \rfloor + \frac{2}{k} \left(k - 2^{\lfloor \log k \rfloor} \right)$$

Proof.
Let f be a minimal code and the codelengths of any two letters, say, the first and the second, differ by more than 1:

$$|f(A_1)| - |f(A_2)| \geq 2. \tag{2.6.1}$$

Pass 1 from the first codelength to the second one, i.e., consider the set of integers

$$|f(A_1)| - 1, \quad |f(A_2)| + 1, \quad |f(A_3)|, \ldots, |f(A_k)|.$$

Compare the Kraft sums for the old and the new sets

$$\left(2^{-|f(A_1)|} + 2^{-|f(A_2)|} + \sum_{i=3}^{k} 2^{-|f(A_i)|} \right) - \left(2^{-|f(A_1)|+1} + 2^{-|f(A_2)|-1} + \sum_{i=3}^{k} 2^{-|f(A_i)|} \right) =$$

$$= -2^{-|f(A_1)|} + \frac{1}{2} 2^{-|f(A_2)|} = 2^{-|f(A_2)|} \left(\frac{1}{2} - 2^{|f(A_2)|-|f(A_1)|} \right) \tag{2.6.2}$$

We get from (2.6.1) and (2.6.2) that the difference between the Kraft sums for the old and the new sets is nonnegative, i.e., the new set meets the Kraft inequality. Hence, it is a set of codelengths of a code. Repeat the pass procedure as many times as it is possible. We get a code, whose codelengths may differ by not more than 1. The procedure does not change the cost. Let a word of the obtained code has got length $l - 1$, $k - 1$ words have got the length l, $0 \leq a < k$, $l \geq 1$. Then the cost equals

$$\frac{1}{k}(kl - a),$$

the Kraft inequality takes the shape

$$k + a \leq 2^l. \tag{2.6.3}$$

From (2.6.3) we obtain that

$$l \geq \lfloor \log k \rfloor \tag{2.6.4}$$

The inequality (2.6.4) turns to the equality iff k is a power of two. Then $a = 0$, and the Claim is proven. If

$$l \leq \lfloor \log k \rfloor + 1,$$

then

$$a = 2^{\lfloor \log k \rfloor + 1} - k$$

yields the optimal code. If

$$l \geq \lfloor \log k \rfloor + 2,$$

then the code is not optimal, Q.E.D.

If one wants to represent decimal digits by binary words, then

$$k = 10, \quad \lfloor \log k \rfloor = 3, \quad a = 6, \quad l = 4.$$

Six digits must have codes of the length 3, four digits must have codes of the length 4. The mapping

$$0 \rightarrow 000, \quad 1 \rightarrow 001, \quad 2 \rightarrow 010, \quad 3 \rightarrow 011,$$

$$4 \rightarrow 100, \quad 5 \rightarrow 101, \quad 6 \rightarrow 1100, \quad 7 \rightarrow 1101, \quad 8 \rightarrow 1110, \quad 9 \rightarrow 1111.$$

is optimal.

Sources of equiprobable letters are closely related to boolean threshold functions, which, in their turn, are used in information retrieval.

Let N and p be natural, $Th_p : E^n \rightarrow E^1$ be the map, which takes to 1 all binary N-length words with p or more ones. The number p is the threshold of that map. We can write

$$Th_p(x_1, \ldots, x_N) = \vee x_{i_1} \ldots x_{i_p} \tag{2.6.5}$$

The disjunction on the right side is taken over all sets, which consist of p variables from N boolean variables $x_1, \ldots, x_N$. We can also write

$$Th_p(x_1, \ldots, x_N) = x_1 Th_{p-1}(x_2, \ldots, x_N) \vee \bar{x}_1 Th_p(x_2, \ldots, x_N) \tag{2.6.6}$$

The complexity of a formula is the quantity of its letters. So, the complexity of formula (2.6.5) is pC_N^p.

The symbol $\lambda_p(N)$ stands for the minimal complexity of formulas representing $Th_p(N)$.

Claim 2.6.2.
Let $2 \leq p \leq N - 2$. *the complexity* $\lambda_p(N)$ *of formulas in the basis: disjunction* $\vee$, *conjunction* $\cdot$, *negation – cannot be less than* $CN \log N$, $C = \text{const}$:

$$\lambda_p(N) \geq CN \log N.$$

Proof.
Take the threshold 2 function first:

$$Th_2(x_1, \ldots, x_n) = \vee_{1 \leq ij \leq N} x_i x_j.$$

The disjunction of a nonempty set of variables is called a type 1 formula. The conjunction of two type 1 formulas F_1 and F_2 is a type 2 formula F, if none of variables of F_1 is a variable of F_2. The formulas F_1 and F_2 are the left and right multipliers of F. The disjunction of m, $m \geq 1$, type 2 formulas is a type 3 formula. The number m is the quantity of disjunctive items of F_2. There is an example of a type 3 formula here:

$$x_1 x_2 \vee (x_1 \vee x_2) x_3 \vee (x_4 \vee x_5)(x_1 \vee x_3). \tag{2.6.7}$$

As it is proven in Krichevskii (1964), each threshold function has a type 3 minimal formula. It is hard to prove that there is a minimal formula, which does not use the negation.
So, take a type 3 formula for $Th_p(x_1, \ldots, x_N)$. It can be described by a matrix with N columns. The number of its rows equals the member of disjunctive items of the formula. At the intersection of the i-th row and the j-th column there is:

- 0, if the j-th variable belongs to the left multiplier of the i-th disjunctive item

- 1, if it belongs to the right one

- α, if it does not belong to any of them

Formula (2.6.7) is described by the matrix

$$\begin{array}{ccccc} 0 & 1 & \alpha & \alpha & \alpha \\ 0 & 0 & 1 & \alpha & \alpha \\ 1 & \alpha & 1 & 0 & 0 \end{array}$$

Let m be the number of disjunctive items in an optimal formula for $Th_p(x_1 \ldots, x_N)$; the number of times the variable x_i enters the formula be l_i, $i = 1, \ldots, N$. There are $d - l_i$ symbols α in the i-th column of the matrix. Change any symbol α in it for 0 or 1. Then the ith column generates 2^{d-l_i} binary columns consisting of 0 and 1. The formula represents the function $Th_2(x_1 \ldots, x_N)$. Hence, for any two variables there is a disjunctive item such that one variable belongs to its left multiplier, the other - to the right one. Consequently, for any two columns of the matrix there is a row, which has 1 at its intersection with one column and 0 at its intersection with the other. Hence, all binary columns generated by the columns of the matrix are different. Their total quantity can not exceed 2^d :

$$\sum_{i=1}^{N} 2^{d-l_i} \leq 2^d.$$

Divide by 2^d both sides. We obtain the known Kraft inequality

$$\sum_{i=1}^{N} 2^{-l_i} \leq 1.$$

The complexity $\lambda_2(N)$ coincides with the cost of an optimal prefix code:

$$\lambda_2(N) = \sum_{i=1}^{N} l_i$$

By Claim 2.3.1,
$$\lambda_2(N) \geq N \log N. \tag{2.6.8}$$

Using the relation
$$\overline{Th_p}(\bar{x}_1, \ldots, \bar{x}_N) = Th_p(x_1, \ldots, x_N)$$

we can obtain an optimal formula for Th_p, where $\frac{N}{2} \leq p \leq N$, from such a formula for $p \leq \frac{N}{2}$. So, let p be an integer, $\frac{N}{2} \leq p \leq N$. Take a formula for $Th_p(x_1, \ldots, x_N$ and let $x_1 = \ldots = x_{p-2} = 1$. By (2.6.6) we get a formula for $Th_2(x_{p-1}, \ldots, x_N)$. The complexity of the formula for $Th_2(x_{p-1}, \ldots, x_N)$ can not exceed the complexity of the formula for $Th_p(x_1, \ldots, x_N)$:

$$\lambda_p(N) \geq \lambda_2(N - p + 2) \tag{2.6.9}$$

From (2.6.8), (2.6.9) and the inequality $p \geq \frac{N}{2}$ we get

$$\lambda_p(N) \geq \lambda_2 \left(\frac{N}{2} + 2 \right) \geq \frac{1}{2} N \log \frac{N}{2},$$

Q.E.D.

Claim 2.6.3.
The complexity of a minimal formula for $Th_2(x_1, \ldots, x_N)$ equals

$$N \lfloor \log N \rfloor + 2 \left(N - 2^{\lfloor \log N \rfloor} \right)$$

Proof.
It follows from the correspondence between the sources with equiprobable letters and formulas for $Th_2(x_1, \ldots, x_N)$. The correspondence was set in the proof of Claim 2.6.2.

2.7 Stationary Sources

The cost of BV-encoding of a stationary source is defined by (2.1.11). For the redundancy $R(f, S)$ of such an encoding on a source S we have

$$R(f, S) = \frac{1}{n} \sum_{x, x \in A^n} p(x) |f(x)| - H(S), \qquad (2.7.1)$$

where A is an alphabet, $H(S)$ us the entropy of a source, n is the blocklength.

Claim 2.7.1.
The redundancy of any prefix block encoding of any stationary stochastic source is nonnegative. There is a prefix block encoding with arbitrary small redundancy. The redundancy of any prefix encoding of any Bernoulli source does not exceed $\frac{1}{n}$, n being the blocklength.

Proof.
The n-length words are the atoms of the partition $S^1 \vee S^2 \vee \ldots S^n$ defined in 1.4. They may be considered as letters of a finite stochastic source (superletters). A map f being prefix, we obtain by Claim 2.3.1, that

$$\sum_{x, |x| = n} p(x) |f(x)| - H(S^1 \vee \ldots \vee S^n) \geq 0 \qquad (2.7.2)$$

For the Shannon encoding we get

$$1 \geq \sum_{x, |x| = n} p(x) |f(x)| - H(S^1 \vee \ldots \vee S^n) \qquad (2.7.3)$$

Relations (2.7.1) - (2.7.3) yield:
$$C(f, S) \geq 0$$

For the Shannon encoding

$$R(f, S) \leq \frac{1}{n} + \frac{1}{n} \left(H(S^1 \vee \ldots \vee S^n) - H(S) \right)$$

Hence, the per letter redundancy of the Shannon encoding goes to zero as the blocklength goes to infinity. For the Bernoulli source it does not exceed $\frac{1}{n}$, Q.E.D.

We want to find a relation between $R(f, S)$ and the block-length n. As a rule,

$$\frac{c_1}{n} \leq R(f, S) \leq \frac{c}{n}$$

The only exception is the Bernoulli source with concerted probabilities. All numbers $\log \frac{p(A_i)}{p(A_j)}$, $1 \le i, j \le k$ for such a source are integers.

First, we prove a Lemma of G. L. Khodak. The symbol $\|x\|$ stands for the distance from a number x to the nearest integer.

Lemma 2.7.1.
Let S be a finite stochastic source, $S = \{A_1, \ldots, A_k\}$, the probability of A_i be $P(A_i) = P_i$, $i = 1, \ldots, k$, $k \ge 1$, f be a prefix code, $R(f, S)$ be its redundancy.
Then

$$R(f, S) \ge \frac{\ln 2}{2\sqrt{2}} \sum_{i=1}^{k} P_i \left\| \log \frac{1}{P_i} \right\|^2$$

Proof.
Let

$$\hat{x} = \begin{cases} -1/2, & x \le -1/2 \\ x, & |x| \le 1/2 \\ 1/2, & x > 1/2 \end{cases}$$

Then

$$|\hat{x}| = \|x\|, \quad \text{if } |x| < 1/2,$$
$$|x| = \tfrac{1}{2} \ge \|x\|, \quad \text{if } |x| > 1/2.$$

Thus,

$$\hat{x}_i^2 \ge \|x_i\|^2 \tag{2.7.4}$$

where

$$x_i = |f(A_i)| + \log p(A_i),$$

f is a prefix code. The codelength $|f(A_i)|$ is an integer, hence

$$\|x_i\| = \left\| \log \frac{1}{p(A_i)} \right\|$$

and (2.7.4) can be rewritten as

$$\hat{x}_i^2 \ge \| -\log p(A_i) \|, \quad i = 1, \ldots, k, \tag{2.7.5}$$

Introduce a function $\lambda(x)$ by

$$\lambda(x) = 2^{-x} + x \ln 2 - 1$$

Its first and second derivatives are

$$\lambda'(x) = \ln 2(1 - 2^{-x}), \quad \lambda''(x) = \ln{}^2 2 \cdot 2^{-x}.$$

The first derivative is positive, if $x > 0$. It is negative, if $x < 0$. Combining it with the inequalities

$$x \ge \hat{x}, (x > 0); \quad \text{and } x \le \hat{x}, (x < 0)$$

we obtain

$$\lambda(x) \geq \lambda(\hat{x}) \tag{2.7.6}$$

Take the Taylor expansion for $\lambda(\hat{x})$ with the remainder in the Lagrange form:

$$\lambda(\hat{x}) = \frac{1}{2} \ln {}^2 2 \cdot 2^{-\xi} \hat{x}^2, \tag{2.7.7}$$

where $|\xi| < \frac{1}{2}$, because $|\hat{x}| \leq \frac{1}{2}$. This expansion and (2.7.6) yield

$$\lambda(x) \geq \frac{1}{2} \ln {}^2 2 \cdot \frac{1}{\sqrt{2}} \hat{x}^2 \tag{2.7.8}$$

Use the Kraft inequality

$$0 \geq 1 - \sum_{i=1}^{k} 2^{-|f(A_i)|}$$

and the definitions of x_i and $\lambda(x)$. We get the following chain:

$$0 \geq 1 - \sum_{i=1}^{k} 2^{-|f(A_i)|} = 1 - \sum_{i=1}^{k} p(A_i) 2^{-x_i} =$$

$$1 - \sum_{i=1}^{k} (1 - x_i \ln 2 + \lambda(x_i)) p(A_i) = \ln 2 \sum_{i=1}^{k} x_i p(A_i) - \sum_{i=1}^{k} p(A_i) \lambda(x_i) \tag{2.7.9}$$

Applying to (2.7.9) inequality (2.7.8) and the definition of the redundancy yield

$$0 \geq \ln 2 \cdot R(f, S) - \frac{1}{2\sqrt{2}} \ln {}^2 2 \cdot \sum_{i=1}^{k} \hat{x}_i^2 p(A_i) \tag{2.7.10}$$

From (2.7.10) and (2.7.5) we get to the conclusion:

$$R(f, S) \geq \frac{\ln 2}{2\sqrt{2}} \sum_{i=1}^{k} \left\| \log \frac{1}{p_i} \right\|^2$$

Q.E.D.

Theorem 2.7.1. *The per letter redundancy of any Markov finite order source but the Bernoulli one with concerted probabilities is not less than $\frac{c}{n}$, $c > 0$:*

$$R(f, S) \geq \frac{c}{n}$$

For the Bernoulli source with concerted probabilities

$$\lim_{n \to \infty} nR(f, S) = 0.$$

Proof.
1. Let S be a finite order Markov source but Bernoulli one. By Claim 1.4.3

$$H(S^1 \vee \ldots \vee S^n) \geq nH(S) + c$$

The statement of the Theorem follows from this inequality and the definitions of the cost and the entropy.
2. Next let S be a Bernoulli source. We get

$$\sum_{x,\ |x|=n} p(x) \log \frac{1}{p(x)} = H(S^1 \vee \ldots \vee S^n) = nH(S).$$

By Lemma 2.7.1 we have for any prefix code f

$$nR(f,S) = \sum_{x,\ |x|=n} p(x) \left(|f(x)| + \log p(x)\right) \geq$$

$$\frac{\ln 2}{2\sqrt{2}} \sum_{x,\ |x|=n} p(x) \left\| \log \frac{1}{p(x)} \right\|^2 \tag{2.7.11}$$

Let

$$P = (P(A_1), \ldots, P(A_k)) = (P_1, \ldots, P_k)$$

be a k-dimensional probability vector,

$$r = r(x) = (r_1(x), \ldots, r_k(x)) = (r_1, \ldots, r_k)$$

be the vector, whose i-th coordinate equals the number of occurences of the letter A_i in the word x, $i = 1, \ldots, k$. S being a Bernoulli source, we have

$$P = P^r = \left(P(A_1)^{r_1(x)} \cdot \ldots \cdot P(A_k)^{r_k(x)} \right) \tag{2.7.12}$$

The function $\|x\|^2$ is continuous, piecewise differentiable, periodic. Its Fourier expansion

$$\|x\|^2 = \sum_{l=-\infty}^{\infty} \gamma_l e^{2\pi i l x} \tag{2.7.13}$$

converges to it uniformly and absolutely. The numbers γ_l are its Fourier coefficients. Calculate γ_0 :

$$\gamma_0 = \int_0^1 \|x\|^2\, dx = 2 \int_0^{1/2} x^2 dx = \frac{1}{12}$$

Substitute (2.7.12) and (2.7.13) into (2.7.11). Change the order of summation:

$$nR(f,S) \geq \frac{\ln 2}{2\sqrt{2}} \sum_{l=-\infty}^{\infty} \gamma_l \sum_{r,|r|=n} \frac{n!}{r!} \left(P(a_1) e^{2\pi i \log \frac{1}{p_1}} \right)^{r_1} \dots$$

$$\dots \left(p(A_k) e^{2\pi i \log \frac{1}{p_k}} \right)^{r_k} = \frac{\ln 2}{2\sqrt{2}} \sum_{l=-\infty}^{\infty} \gamma_l \left(p_1 e^{2\pi i \log \frac{1}{p_1}} + \dots + p_k e^{2\pi i \log \frac{1}{p_k}} \right)^n. \quad (2.7.14)$$

By the triangle inequality

$$\left| p_1 e^{2\pi i \log \frac{1}{p_1}} + \dots + p_k e^{2\pi i \log \frac{1}{p_k}} \right| \leq 1. \quad (2.7.15)$$

Two cases may present themselves: either (2.7.15) is a strict inequality, or it turns to an equality for some l.

Go to the limit in (2.7.14) in the first case. It is allowed by the uniform convergence. We get

$$nR(f,S) \geq \frac{\ln 2}{2\sqrt{2}} \cdot \frac{1}{12}$$

and the Theorem is proven in that case.
In the second case the arguments of all addends in (2.7.15) are equal, i.e.,

$$\log \frac{1}{p_i} = a \log \frac{1}{p_j} + b$$

$1 \leq i, j \leq k$, a and b are integers. If $a = 1$ or $b = 0$, we get a Bernoulli source with concerted probabilities. We discuss it later. Just now let $a \geq 2$, $b \neq 0$.

Let

$$\psi(l) = 1, \quad l \equiv 0 \bmod a, \quad \psi(l) = 0, \quad l \not\equiv 0 \bmod a,$$

l be an integer. The function φ be an integer. The function φ is represented by its discrete Fourier transformation:

$$\psi(l) = \frac{1}{a} \sum_{v=0}^{a-1} e^{\frac{2\pi i v l}{a}}. \quad (2.7.16)$$

Let n go to infinity in (2.7.14). The limit of all but those with $l \equiv 0$ addends is zero. Inequality (2.7.14) takes the form

$$\lim_{n \to \infty} n \cdot R(f,S) \geq \frac{\ln 2}{2\sqrt{2}} \sum_{l=-\infty}^{\infty} \gamma_l \psi(l) e^{2\pi i l \log \frac{1}{p_1}}, \quad (2.7.17)$$

Substitute (2.7.16) into (2.7.17) and change the summation order:

$$\lim_{n \to \infty} n \cdot R(f,S) \geq \frac{\ln 2}{2a\sqrt{2}} \sum_{v=0}^{a-1} \sum_{l=-\infty}^{\infty} \gamma_l e^{2\pi i l \left(\frac{v}{a} + \log \frac{1}{p_1} \right)}. \quad (2.7.18)$$

Inequality (2.7.18) and expansion (2.7.13) yeild

$$\lim_{n\to\infty} n \cdot R(f,S) \geq \frac{\ln 2}{2a\sqrt{2}} \sum_{v=0}^{a-1} \left\| \log \frac{1}{p_1} + \frac{v}{a} \right\|^2. \tag{2.7.19}$$

Consider the sequence

$$\left\{ \frac{1}{a} \sum_{v=0}^{a-1} \left\| \xi + \frac{v}{a} \right\|^2 \right\}, \quad a = 2, 3, \ldots,; \quad 0 \leq \xi \leq 1.$$

It consists of the Riemann integral sums for $\int_0^1 \|x\|^2 \, dx$. Thus, this integral is the limit of the sequence. Consequently, if $a \geq 2$, then

$$\inf_{\xi, 0 \leq \xi \leq 1} \frac{1}{a} \sum_{v=0}^{a-1} \left\| \xi + \frac{v}{a} \right\|^2 \geq C \geq 0$$

and

$$\inf_{a \geq 2} \inf_{\xi, 0 \leq \xi \leq 1} \frac{1}{a} \sum_{v=0}^{a-1} \left\| \xi + \frac{v}{a} \right\|^2 > C > 0.$$

Go back to (2.7.19):

$$\lim_{n\to\infty} nR(f,S) \geq C > 0.$$

The Theorem is proven in that case.
3. It remains to consider the case of Bernoulli source with concerted probabilities. Then for a sequence $\{n_1, n_2, \ldots\}$

$$\lim n_i R(f,S) = 0.$$

One has to represent $\log \frac{1}{p_1}$ as a continued fraction. If n_i is a denominator of a convergent of the fraction, then

$$n_i R(f,S) < \frac{1}{n_i}.$$

Q.E.D.

G. L. Khodak (1969) discovered a simple variable length to block code. Its redundancy decreases nearly as fast as one of BV-code, i.e., as $\frac{c}{d}$, d being the average coding delay. Khodak's code was later described independently by F. Jelinek and K. Shneider [1972].

Claim 2.7.2. (G. L. Khodak)
For any Bernoulli source there is a VB-code Kh, whose redundancy does not exceed $\frac{c}{d}$:

$$R(Kh, S) \leq \frac{c}{d}$$

where d is the average height of the coding tree (coding delay).

Proof.

The letters $\{A_1, \ldots, A_k\}$ of a source S are supposed to be ordered by their probabilities:

$$p(A_1) \geq \ldots \geq p(A_k).$$

Fix a number m, $m \to \infty$. Construct inductively a tree Δ. Make a word x to be a leaf of Δ, if minimum logarithm of the probability does not exceed m, whereas minus logarithm of the probability of one of its sons does:

$$-\log p(x) \leq m \tag{2.7.20}$$

$$-\log p(A_k)p(x) > m \tag{2.7.21}$$

From (2.7.20) we have that

$$p(x) \geq \frac{1}{2^m}.$$

It means hat the number of leaves of the tree is not more than 2^m. Hence, each leaf can be provided with an m-length binary word. So, a VB-code is defined. Use Claim 2.1.1 and the relation

$$- \sum_{x \text{ is a leaf of } \Delta} p(x) \log p(x) = dH(S)$$

of Lemma 2.1.1. We obtain

$$R(Kh, S) = \frac{1}{d}(m - H(S)) \leq \frac{\log p(A_k)}{d},$$

Q.E.D.

We make some remarks on VV-coding. As it is easily seen from Claim 2.1.1 and Lemma 2.1.1, the redundancy of such a code is nonnegative. G. L. Khodak constructed a VV-code for Bernoulli sources. Its redundancy is not greater than $cd^{-5/3}$.

A commonly used run length encoding is a VV-code. The alphabet is binary. The leaves of the coding tree are $1, 01, 001, \ldots$. In other words, one waits for the first unit to appear. The length of the preceding run of zeroes is encoded and transmitted. The coding tree is infinite. To make it finite one can wait not more than a fixed time T.

The length of runs may be encoded by the Huffman code. Shannon proposed the following code. A codeword is chosen for the less probable letter, say, 1. This word plays the role of comma. To encode lengths of runs the binary words are taken in increasing order, albeit the comma is omitted. If the probability of one goes to zero and the length of the comma goes to infinity, then the redundancy of such a coding goes to zero.

Concluding the Section we discuss the encoding of stationary combinatorial sources (Section 1.5). It is natural to use BB-codes for such sources. If a block-to-block code f maps n-length words to l-length ones, then its redundancy on a source S is

$$R(f, S) = \frac{l}{n} - h(S),$$

where $h(S)$ is the limit per letter entropy of S.

Claim 2.7.3.
The redundancy of any BB-code on any stationary combinatorial source is nonnegative. For any $\varepsilon > 0$ and any source S there is a BB-code f, for which $R(f, S) < \varepsilon$.

Proof.
There are two kinds of stationary combinatorial sources:$S \leq A^*$ or $S \subseteq A^\infty$, A being the alphabet. If $h(n)$ is the logarithm of either the number of n-length prefixes ($S \subseteq A^\infty$), or the number of n-length words ($S \subseteq A^*$), then

$$h(S) \leq \frac{h(n)}{n}, \quad h(S) = \lim_{n \to \infty} \frac{h(n)}{n}$$

(Claim 1.5.2). Every n-length word w is one-to-one represented by its l-length code. Hence,

$$2^l \geq 2^{nh(n)} \geq 2^{nh(S)},$$

and the redundancy is proven to be nonnegative.

For any $\varepsilon > 0$ and a big enough n the number of either n-length words or prefixes in S is not more than

$$2^{n(h(S)+\varepsilon/2)}.$$

If

$$l = \lceil n(h(S) + \varepsilon/2) \rceil,$$

then there is an injective map f of the set of n-length words of S to the set of l-length binary words.

The redundancy $R(f, S)$ of such a map is not more than ε, $n \to \infty$. Q.E.D.

Combine Claim 2.7.3 with Claim 1.5.3. We get

Corollary 2.7.4.
Let S be a stationary combinatorial source over an alphabet A, $S \subseteq A^*$, S does not coincide with A^*. Then for any $\varepsilon > 0$ there is a BB-code f, mapping n-length words to l-length ones, whose cost $\frac{l}{n}$ is less than $\log_2 |A|$.

2.8 Kolmogorov Complexity

If a word x is generated by a combinatorial source S, then it is given a code, whose length is $H(S)$. The word can be reconstructed by its code. The codelength depends on the source. It would be quite desirable to provide any word with an invariant code independent of any source. Such an absolute code was invented by A.N.Kolmogorov.

Everybody would agree, that the word $010101\ldots$ is more complex, than $000000\ldots$ A word corresponding to coin throwing is usually more complex than both of those words. The computer program which generates $000\ldots$ is rather simple: print 0 from the beginning to the end. The program which generates $0101\ldots$ is just a bit more complicated: print the symbol opposite to one printed at the previous moment. A chaotic word can not be produced by any short program.

Suppose we have got a computer $\mathbf{C}$ supplied with a program P. Being fed with a word x the computer $\mathbf{C}$ produces a word y. The word x is called a code of y with respect to P. The minimal length of all codes of y with respect to P is called the complexity of y with respect to P. It is denoted by $K_P(y)$. If there is no such a word, then $K_P(y) = \infty$.

The computer $\mathbf{C}$ is fixed. The complexity $K_P(y)$. depends on the program P. It is very important, that there is a complexity which is program-independent to within an additive constant.

Claim 2.8.1. (Universal program)
There is a universal program P_0. The complexity of any word with respect to P_0 is not greater than its complexity with respect to any other program to within an additive constant. For any program P there is a positive constant such that for each y

$$K_{P_0}(y) \leq K_P(y) + \mathbf{C}$$

Proof.
Introduce first a convenient notation, which is called doubling. If $x = x_1,\ldots,x_n$ is a binary word, then $\bar{x}$ is the word obtained by doubling the letters of x and appending 01 at the end:

$$\bar{x} = x_1 x_1 \ldots x_n x_n 01.$$

Using $\bar{x}$, one can separate words without commas: given $\bar{x}y$, one can find both x and y.

Address the theory of recursive functions. An important statement of the theory says, that there is a function $\mathbf{C}_0$ universal for the set of partially recursive functions. It means that every partially recursive function $\mathbf{C}$ has got a number. Provided with that number, $\mathbf{C}_0$ maps a word x to the word y, which $\mathbf{C}$ maps x to, i.e., $\mathbf{C}_0$ imitates every partially recursive function $\mathbf{C}$.

We take for granted a computer analogue of that statement. It says, that there is a computer $\mathbf{C}_0$ universal for the set of computers. Every computer $\mathbf{C}$ has got a detailed description. Provided with both that description and a program P, $\mathbf{C}_0$ maps a word x to the word y, which $\mathbf{C}$, being loaded with P, maps x to. We can say that $\mathbf{C}_0$ imitates the work of every computer $\mathbf{C}$.

Suppose that a computer $\mathbf{C}$ was chosen. Let x be a code of a word y with respect to a program P, i.e., $\mathbf{C}$, being loaded with P, produces y from x. Change the computer

C for the universal computer $\mathbf{C}_0$. Define a code of y to be $\bar{P}x$, its length to be

$$2\,|P| + 2 + |x|\,.$$

Define a program P_0. Given a word $\bar{P}x$, it first separates both P and x. Then, $\mathbf{C}_0$ being a universal computer, it maps x to y. Hence, the codelength of y with respect to P_0 is

$$2\,|P| + 2 + |x| = |x| + \text{const}\,.$$

In other words, we have developed a program P_0. For any program P there is a constant C such that

$$K_{P_0}(y) \le K_P(y) + C,$$

for every y. That constant is $2\,|P| + C$, Q.E.D.

Chose any universal program P_0. The complexity of a word x with respect to P_0 is called the Kolmogorov complexity of x and denoted by $K(x)$. Changing the program P_0, we get $K(x) + C$ instead of $K(x)$ for all x. Asymptotical properties of $K(x)$ do not depend on the choice of any particular universal program.

The less the Kolmogorov complexity of a word x, the shorter its code. A word with small complexity is easy to remember and reproduce. It is worth noting that even ants do have some idea of Kolmogorov complexity. E.g., the word 000000 takes less time to be conveyed by ants-explorers than the word 010010, see section 1.1.

The map $x \to K(x)$ is quite exotic.

It is uncomputable. In order to get some idea of its behaviour, we introduce two functions:

$$\bar{K}(n) = \max_{|x|=n} K(x)$$

and

$$\underline{K}(n) = \min_{|x|=n} K(x).$$

An universal program P_0 is fixed.

Claim 2.8.2. (Maximal and minimal Kolmogorov complexities of n-length words) *Let $\bar{K}(n)$ be the maximal Kolmogorov complexity of n-length binary words, $\underline{K}(n)$ be the minimal one, $n \ge 1$. Then*
  1.
$$n \le \bar{K}(n) \le n + C, \qquad C = \text{const}$$

For any $\varepsilon > 0$ only a vanishing part of all binary n-length words have complexity less than $n(1 - \varepsilon)$, $n \to \infty$
  2.
$$\overline{\lim_{n \to \infty}} \frac{\underline{K}(n)}{\log n} = 1$$

  3. *For any arbitrary slow increasing computable function $\varphi(n)$*

$$\lim_{n \to \infty} \frac{\underline{K}(n)}{\varphi(n)} = 0.$$

Proof.
1. A word x, $|x| = n$, is given the code x by the identical map P. So,

$$K_P(x) = n, \qquad K_{P_0}(x) \le n + C.$$

The rest of the first statement is a rephrasing of Claim 2.3.2, where $|S| = 2^n$, $t = \varepsilon n$.
2. If a word x consists of zeros, encode it by 1 concatenated with Bin $|x|$:

$$0 \to 11, \quad 00 \to 110, \quad 000 \to 111, \quad 0000 \to 1100, \quad \text{etc.}$$

If there is a one in x, encode it by $1x$. Let P be a corresponding decoding program. We get

$$\underline{K}(n) \le K(\underbrace{0\ldots0}_{n \text{ times}}) = K_{P_0}(0\ldots0) \le$$

$$\le K_P(0\ldots0) + C \le \log n + C. \tag{2.8.1}$$

From (2.8.1):

$$\overline{\lim_{n\to\infty}} \frac{K(n)}{\log n} \le 1. \tag{2.8.2}$$

Suppose that

$$\overline{\lim_{n\to\infty}} \frac{K(n)}{\log n} < 1.$$

Then there are $\alpha < 1$ and n_0 such that for every $n > n_0$

$$\underline{K}(n) < \alpha \log n.$$

So, there are $n - n_0$ words $x_{n_0}, \ldots, x_n$, where $|x_i| = i$, $n_0 < i \le n$,

$$K(x_i) \le \alpha \log n. \tag{2.8.3}$$

For each x_i there is a code, whose length $K(x_i)$ meets (2.8.3). The quantity of such codes is not greater than $2 \cdot n^\alpha$. Thus,

$$n - n_0 \le 2n^\alpha, \qquad \alpha < 1, \quad n \to \infty \quad -$$

a contradiction. The second statement is proven.
3. Let $\varphi(n)$ be a monotone computable function and

$$\lim_{n\to\infty} \frac{K(n)}{\psi(n)} = \alpha > 0.$$

There is a sequence $n_1, n_2, \ldots$ such that

$$\underline{K}(n_i) > (\alpha - \varepsilon)\psi(n_i) \tag{2.8.4}$$

where $\alpha - \varepsilon > 0$, $i = 1, 2, \ldots$.

Take a positive integer m. Compute $\psi(1), \psi(2), \dots$. Since ψ is an increasing function, there is a number n_j such that

$$m \leq (\alpha - \varepsilon)\psi(n_j). \qquad (2.8.5)$$

Take any n_j-length word x, $x = x(m)$. We have just defined a program P, which maps every integer m to a word $x(m)$. We have

$$K_P(x(m)) \leq \log m + 1, \qquad (2.8.6)$$

since the codelength of $x(m)$ with respect to P is

$$|\text{Bin } m| \leq \log m + 1.$$

Let P_0 be an universal program. Then

$$K_{P_0}(x(m)) \leq C + K_P(x(m)). \qquad (2.8.7)$$

By the definition of $\underline{K}(n)$,

$$\underline{K}(n_j) \leq K_{P_0}(x(m)). \qquad (2.8.8)$$

From (2.8.4)-(2.8.8) we obtain:

$$m \leq (\alpha - \varepsilon)\psi(n_j) < \underline{K}(n_j) \leq K_{P_0}(x_m) \leq$$

$$\leq C + K_P(x_m) \leq \log m - C$$

— a contradiction, since $m \to \infty$. Q.E.D.

Corollary 2.8.1.
It is impossible to compute the map $x \to K(x)$, where $K(x)$ is the Kolmogorov complexity of x.

Proof.
Suppose one can compute $K(x)$ for every x. Then one can find $\underline{K}(n)$ for every n and

$$\lim_{n \to \infty} \frac{K(n)}{\varphi(n)} = 1,$$

where $\varphi(n) = \underline{K}(n)$ is a computable function. It contradicts Claim 2.8.2. Q.E.D.

2.9 Majorizing the Kolmogorov Complexity

The Kolmogorov complexity is the highest degree to which a word can be condensed. It is the best solution of the data compression problem. The only drawback is its uncomputability. Computable substitutes of the Kolmogorov complexity are used in practice. They majorize the complexity.

A function $f(x)$ is a majorant of the Kolmogorov complexity, if there is $C > 0$ such that for any word x

$$K(x) \leq f(x) + C \qquad (2.9.1)$$

First we give two general claims on the majorants due to A. N. Kolmogorov and L. A. Levin. Then we describe some of them.

Claim 2.9.1.
There is a computable function $K(t, x)$ such that for any word x

$$\lim_{t \to \infty} K(t, x) = K(x).$$

If $t_1 \leq t_2$, then

$$K(t_2, x) \geq K(t_1, x) \geq K(x).$$

Proof.
Take a constant C, $C > 0$, such that for every x

$$K(x) \leq |x| + C,$$

a universal program P_0 and a computer $\mathbf{C}_0$. Given a word x and a number t, let $\mathbf{C}_0$ work t units of time on all binary words w, $|w| \leq x + C$. If $\mathbf{C}_0$ produces the word x, let $K(t, x)$ equal the minimal length of those w, which give x. If it do not produces x, let

$$K(t, x) = |x| + C.$$

It is clear, that $K(t, x)$ is computable and

$$K(t_2, x) \geq K(t_1, x) \geq K(x).$$

Letting t go to infinity, we find sooner or later a word w, such that $|w| = K(x)$, Q.E.D.

We can never be sure, that the next step will not give for x a better program.

So, $K(t, x)$ is a family of majorants.

Claim 2.9.2. (Criterion of majorizing)
A computable function f is a majorant of the Kolmogorov complexity, iff for any integer a there is $C > 0$ such that

$$\log |\{x : f(x) = a\}| \leq a + C.$$

Here $|\{x : f(x) = a\}|$ is the quantity of words x such that $f(x) = a$.
Proof.
Let f be a majorant. Inequality (2.9.1) is met. If $f(x) = a$, then

$$K(x) \leq a + C \qquad (2.9.2)$$

The quantity of words x, for which (2.9.2) holds, is not greater than 2^{a+C+1}. Hence,

$$|\{x : \; f(x) = a\}| \leq 2^{a+C+1},$$

and a part of the Claim is proven.

Let for any a

$$|\{w : \; f(w) = a\}| \leq 2^{a+C}, \tag{2.9.3}$$

C be natural. Take a word x, let $f(x) = a$. Give every word w, for which $f(x) = a$, a distinct number P. From (2.9.3),

$$|P| \leq a + C \tag{2.9.4}$$

The word x is given a number $P = P(x)$ as well. Define $\bar{P}$ as the word obtained by putting before P, first, one, and, second, $a + C + 1 - |P|$ zeroes:

$$\bar{P} = 0\ldots01P, \qquad |\bar{P}| = a + C + 1.$$

Given C, extract $C + 1$ from $|\bar{P}|$. We get a. Omit zeroes and one in $\bar{P}$. We get P. Take the P-th element of the set

$$\{w : \; f(w) = a\}.$$

It is just x. So, $\bar{P}$ is a code of x. The procedure P that has just been described is a decoding program. Hence,

$$K_P(x) = |\bar{P}| = a + C + 1. \tag{2.9.5}$$

If P_0 is a universal program, then

$$K(x) = K_{P_0}(x) \leq K_P(x) + C_1 = a + C_2. \tag{2.9.6}$$

Substitute $a = f(x)$ into (2.8.6). We obtain

$$K(x) \leq f(x) + C_2,$$

i.e., f is a majorant. Q.E.D.

Next we list several majorants of the Kolmogorov complexity. Some of them have already been used. The others will be used later.

1. Identity Code.

The identity map $x \to x$ is used as encoding. The codelength $|x|$ is a majorant:

$$K(x) \leq |x| + C$$

2. Source Code.

A finite combinatorial source S determines an encoding f. If a word x is generated by S, then x is given a code $f(x)$,

$$|f(x)| = \lceil H(S) \rceil + 1.$$

The code $f(x)$ is the concatenation of one with the number of x within the set S. If x does not belong to the set S, then its code $f(x)$ is the concatenation of zero with x itself: $f(x) = 0x$. We have got a majorant of the Kolmogorov complexity:

$$K(x) \le 1 + \log \lceil |S| \rceil + C, \quad x \in S; \qquad K(x) \le |x| + C, \quad x \notin S.$$

If S is an infinite set, then its members can be numbered, $f(x)$ is defined the same way. Such a majorant was used in the proof of Claim 2.8.2 for the set

$$\{0, 00, 000, \ldots\}$$

Codes determined by stationary combinatorial sources were, as a matter of fact, developed in Claim 2.7.3.

A stationary source S, $S \subset A^*$ generates finite words. For any $\varepsilon > 0$ there is a number n_0 such that for every $n \ge n_0$ the quantity of n-length words is

$$2^{n(h(S)+\varepsilon/2))}$$

Fix a number n, $n \ge n_0$.

The list of n-length words is called the vocabulary and denoted by V. Each word w, n_0 being a divisor of $|w|$, is subdivided into n_0-length subwords.

The source S being stationary, the vocabulary V contains each one of them. A subword is given an l-length code,

$$l = \lceil n(h(S) + \varepsilon/2) \rceil.$$

The length of the concatenation of all those codes equals

$$\frac{|w|}{n} \cdot l$$

To reconstruct w, one should know, first, the concatenation, second, the length $|w|$, third, the number $|n|$, fourth, the vocabulary. They are encoded by employing the doubling to avoid commas. So the code of w is

$$\overline{w|\overline{n}} \cdot \overline{V} \cdot \frac{|w|}{n} \cdot l.$$

Its length is

$$2|w| + 2n + 2n \cdot 2^l + 6 + \frac{|w|}{n} \cdot l.$$

If $|w|$ goes to infinity, n and ε are fixed, then the codelength is not greater than

$$|w|\,(h(S) + \varepsilon).$$

Hence, the Kolmogorov complexity of a word w generated by a stationary source S is majorized by

$$|w|\,(h(S) + \varepsilon).$$

$|w|$ is big enough.

If a source S does not generate all words ($S \neq A^*$), then its per letter entropy $h(S)$ is less than $\log_2 |A|$ (Claim 1.5.3). Consider the case of the binary alphabet, $\log_2 |A| = 1$. There are binary words, whose Kolmogorov complexity is not less than their length (Claim 2.8.2, part 1). Such words are most complicated among the words of the same length. We have come to the following conclusion.

Claim 2.9.3.
If at least one word is not produced by a binary stationary source S, then S generates only a finite number of most complicated words.

Apply Claim 2.9.3 to Jablonskii invariant classes. The only invariant class which contains all most complicated functions is the class of all boolean functions.

S. Jablonskii (1959) discussed the problem of generating most complicated boolean functions. He said, that if a function f was generated by an algorithm, then all functions obtained from f by substitution of constants, are generated as well. Consequently, the functions generated by an algorithm constitute an invariant class. If that class includes all most complicated functions, then it must include all boolean functions (Claim 2.9.3). This reasoning may be considered as an argument for the following statement: There is no other way to generate a most complicated function, as to compare its complexity with the complexities of all functions with the same number of variables.

3. Empirical Combinatorial Entropic (ECE) Code.

Such a code was developed in Claim 2.5.2. Given $n > 0$, a word w is broken into $\frac{|w|}{n}$ n-length words. A list V of all different words is called the vocabulary of w. By the definition of empirical entropy, the cardinality $|V|$ of V equals $2^{nCH_n w}$, its size is

$$n\lceil \log |A| \rceil\,|V|.$$

The code of V is $\bar{V}$ (see Claim 2.8.1). We use the doubling $\bar{V}$ to avoid commas. So,

$$|\text{code } V| = 2n\lceil \log |A| \rceil \cdot 2^{nCH_n w}. \tag{2.9.7}$$

Every n-length subword of w is given its number in the vocabulary. The length of the number is $\lceil nCH_n w \rceil$. The concatenation of all those numbers is called the main part of ECE-code. We have

$$|\text{main part}| = \frac{|w|}{n}\lceil nCH_n w \rceil. \tag{2.9.8}$$

To reconstruct w, one should know the length $|w|$ and $|n|$. They are encoded by doubling as well. Hence, ECE-code is defined as the concatenation

ECE(w)=(word length)(the length of n)(code V)(main part).

For the codelength we have from (2.9.7) and (2.9.8):

$$|ECE(w)| = 2\lceil \log |w| \rceil + \lceil 2 \log n \rceil + 2n\lceil \log |A| \rceil 2^{nCH_n w} +$$

$$+6 + \lceil nCH_n w \rceil \cdot \frac{|w|}{n}. \tag{2.9.9}$$

One can find w by ECE(w). The vocabulary V of ECE code consists of equal length words. The codelengths are all equal as well. So, it is a BB-code.

4. Lempel-Ziv Code.
It is a VB-code. A word w is subdivided into subwords, which are determined according to a so-called incremental parsing procedure. This procedure creates a new subword as soon as a prefix of the still unparsed part of w differs from all preceding subwords. There is a number p such that the first p subwords are all distinct from each other and every following subword coincides with one of the first subwords. Those first words make a vocabulary V. Any subword is encoded by its number in V.

There are many of variants of this code. Many of them are good for practical purposes.

5. Huffman and Shannon Empirical Codes.
An n-length subword i of a word w has frequency

$$\frac{nr_i(w)}{|w|}, \qquad i \in A^n.$$

Make either the Shannon or the Huffman code f for that set of frequencies:

$$i \to f(i).$$

There is a table or coding book of f. For any $i \in A^n$ it gives $f(i)$. The concatenation of the codes of all subwords of w prefixed by both the coding book and n in a separable way makes a code of w. It is called the Empirical Shannon or Huffman code of w. Its length is a majorant of the Kolmogorov complexity.

6. Move to Front Code.
Let A be an alphabet, $i \to f(i)$, $1 \le i \le |A|$ be a prefix code. Suppose that the letters of A make a pile, like a pile of books. Let j_1 be the first letter of a word w. Find j_1 in the pile. If it takes the i_1-th position, encode it by $f(i_1)$. Take j_1 out of the pile and put it on the top (move to the front). Repeat the procedure with the second letter of w and the pile obtained, etc. As we will see, the codelength is close enough to $H_A(w)$, if the code f is chosen properly.

7. Empirical Entropic (EE) Code.
Take a word w over an alphabet B. Define a vector

$$r(w) = (r_1(w), \ldots, r_{|B|}(w),$$

$r_i(w)$ being the number of occurences of i in w, $i = 1, \ldots, |B|$. Let $g(r)$ be the set of all words w for which $r(w) = r$. For the cardinality $|g(r)|$ we have

$$|g(r)| = \frac{|r|!}{r!}. \tag{2.9.10}$$

Take the well known Stirling formula

$$\ln \Gamma(a) = \ln \sqrt{2\pi} + \left(a - \frac{1}{2}\right)\ln a - a + \frac{\theta}{12}, \tag{2.9.11}$$

where $\Gamma(a)$ is the Euler gamma function, $\Gamma(a+1) = a!$, if a is an integer, $0 < \theta < 1$. It may be given the form

$$\ln a! = \ln \sqrt{2\pi} + a\ln a + \frac{1}{2}\ln \hat{a} + o\,(1) \tag{2.9.12}$$

where $\hat{a} = \max(a, 1)$.
    Substitute (2.9.12) into (2.9.10):

$$\log|g(r)| = |w|\,H_B(w) - \frac{1}{2}\sum_{i=1}^{|B|}\log\frac{\hat{r}_i}{|w|} - (|B| - 1)\ln \sqrt{2\pi} + \frac{|B| - 1}{2}\log|w| + O\,(1). \tag{2.9.13}$$

A vector

$$r(w) = \big(r_1(w), \ldots, r_{|B|}(w)\big)$$

meets the condition

$$r_1(w) + \cdots + r_{|B|}(w) = |W|. \tag{2.9.14}$$

The quantity of different vectors $r(w)$ equals the number of partitioning $|w|$ into $|B|$ addents, which equals

$$C_{|w|+|B|-1}^{|B|-1}.$$

    Define EE-code of a word w. It consists of a prefix and a suffix. The prefix is the rank of the vector $r(w)$ among all possible vectors. The binary notation of the prefix consists of

$$\log C_{|w|+|B|-1}^{|B|-1} = (|B| - 1)\log|B| + C \qquad \text{bits.} \tag{2.9.15}$$

The suffix is the rank of w in the set $g(r)$, $r = r(w)$. The bitlength of the suffix is given by (2.8.13). Such a code gives a majorant of the Kolmogorov complexity. The main addent of the majorant is $|w|\,H_B(w)$. So,

$$|\mathrm{EE}(x)| = \log|g(r)| + \log C_{|w|+|B|-1}^{|B|-1}.$$

NOTES

The output of a source may be compressed up to its entropy, not more. It is the main fact of the source coding theory, which has its origin in Shannon (1948). The first theoretical compressing code is Shannon's. It happened to be very close to Morse's, which was developed empirically a century earlier. Shannon's code is very simple, although not optimal. An optimal code for a stochastic source was constructed by Huffman (1952). It may be used to compress an individual word. The frequencies play the role of probabilities. The encoding may be either two-pass or one-pass. In the first case the frequencies of letters are counted beforehand, then a code is built. In the second case the code is changed along with the word reading.

Output words of Shannon or Huffman's codes are of different lengths. There is a compressing code, whose output words are of equal length. It was developed by first G. L. Khodak (1969) and then by F. Jelinek and F. Schneider (1972).

A decipherable code for integers is constructed by V. Levenstein (1968) and P. Elias (1975).

G. Hansel (1962) and R. Krichevskii (1963) exploited the source coding to lower-bound the length of threshold formulas.

The Kolmogorov complexity bridges the theory of algorithms and the information theory. Any code, including Lempel-Ziv's, move to front etc., is only a majorant of Kolmogorov's. Sections 2.8 and 2.9 are based on Zvonkin and Levin (1970).

CHAPTER 3

Universal Codes

A universal code is suited for many sources simultaneously. To develop such a code we need only calculate the capacity of a communication channel. Universal codes are constructed explicitly for some interesting sets of sources. Those are Bernoulli, Markov, monotone sources (probabilities of letters are ordered). There are both block-to-variable-length and variable-length-to-block universal codes. There is a sequence of codes, which is weakly universal on the set of all stationary sources (the redundancy goes to zero).

The majorants ot Kolmogorov complexity go to work as universal codes.

An adaptive code depends on the result of observation of a source. We present optimal adaptive codes, which are akin to universal ones.

Universal codes are used to make identifying keys.

3.1 Encoding of Sets of Stochastic Sources

A code is universal, if it is good for several sources. In Chapter 2 we have discussed the compression of data generated by a source S, which was supposed to be known. It means that the probabilities of its letters $p(A_1), \ldots, p(A_k)$ were used to construct a code. Those probabilities remain unchanged.

We get a more realistic setting if suppose that S is known exactly. We know only that S belongs to a set Σ of sources. A good example is the set Σ, which includes English, German and French. All those languages use Latin letters. The probability of a letter depends on the language. Nevertheless, we wish to give each letter a code, which does not depend on the language. We wish to make the code acceptable for any language. It means that the maximal redundancy over the language should be as small as possible.

The redundancy $R(f, \Sigma)$ of a code f on a set Σ is the supremum of its redundancies on sources $S \in \Sigma$:

$$R(f, \Sigma) = \sup_{S \in \Sigma} R(f, S).$$

The sources $S \in \Sigma$ are stochastic. Without loss of generality we suppose that all of them have one and the same alphabet:

$$A = \{A_1, \ldots, A_k\}.$$

A coding map $f : A_i \to f(A_i)$ is identified with the set

$$f = (|f(A_1)|, \ldots, |f(A_k)|)$$

of its codelengths. The map f is prefix. It meets the Kraft inequality

$$\sum_{i=1}^{k} 2^{-|f(A_i)|} \leq 1 \tag{3.1.1}$$

We denote by F the set of all vectors $f = (f_1, \ldots, f_k)$ for which (3.1.1) holds. The coordinates of f are any real numbers. F is convex and closed, since the map f is continuous and convex. If $f = (f_1, \ldots, f_k)$ is a vector, then

$$\lceil f \rceil = (\lceil f_1 \rceil, \ldots, \lceil f_k \rceil) .$$

$\lceil F \rceil$ is the set of all vectors f, such that $f \in F$ and all coordinates of f are integers. It is clear that if $f \in F$, then $\lceil f \rceil \in \lceil F \rceil$. The redundancy of f on S is defined by

$$R(f, S) = \sum_{i=1}^{k} p_S(A_i) \left(\log p_S(A_i) + f_i \right), \tag{3.1.2}$$

where $p_S(A_i)$ is the probability to generate a letter $A_i \in A$ by a source $S \in \Sigma$, $i = 1, \ldots, k$, $k = |A|$, $S \in \Sigma$. As it is easily seen,

$$R(f, S) + 1 \geq R(\lceil f \rceil, S) \geq R(f, S) \tag{3.1.3}$$

The optimal redundancy of encodings of Σ is denoted by $R(\Sigma)$:

$$R(\Sigma) = \inf_{f \in \lceil F \rceil} R(f, \Sigma)$$

A code f, for which $R(f, \Sigma) = R(\Sigma)$, is called optimal for Σ. We want to find an optimal or nearly optimal code for Σ. As we will see, the problem of optimal encoding of a set Σ is equivalent to a well known problem of the information theory: to find the capacity of a communication channel.

Let Q be a probability measure on Σ. Then we get the average redundancy of a code f on Σ with respect to Q :

$$R_Q(f) = \int_{\Sigma} R(f, S) dQ(S)$$

The infimum of $R_Q(f)$ over $f \in F$ is denoted by R_Q.

A set Σ along with a measure Q define a communication channel, see Gallager (1968). Input symbols of the channel are sources $S \in \Sigma$. Output symbols are letters of the alphabet A. As soon as a symbol S appears at the input, then a letter A_i will appear at the output with the probability $P_S(A_i)$, $i = 1, \ldots, k$. The total probability of a letter A_i is

$$p_i = \int p_s(A_i) dQ(S)$$

The information between S and the output is

$$I_Q(S, A) = \sum_{i=1}^{k} p(A_i) \log \frac{p_S(A_i)}{p_i}$$

Averaging $I_Q(S, A)$ we get the information between the input and the output:

$$I_Q(\Sigma, A) = \int I_Q(S, A) dQ(S).$$

The capacity C of the channel is defined as

$$C = \sup_Q I_Q(\Sigma, A),$$

where supremum is taken over all probability measures on Σ. All those definitions go back to Shannon. They can be found in Gallager (1968).

Lemma 3.1.1. (Davisson and Leon-Garcia (1980))
Let A be a finite alphabet, Σ be a set of sources, Q be a measure on Σ, R_Q be the minimal average redundancy, $I_Q(\Sigma, A)$ be the input-output information of the corresponding channel. Then

$$R_Q = I_Q(\Sigma, A)$$

Proof.
We get from (3.1.2) that

$$R_Q(f) = \int_\Sigma R(f, S) dQ(S) = \sum_{i=1}^{k} \int p_s(A_i) \log p_s(A_i) dQ(S) + \sum_{i=1}^{k} f_i p_i \qquad (3.1.4)$$

$R_Q(f)$ takes its minimal value, if

$$f = f^0, \quad f_i^0 = -\log p_i.$$

Substituting f^0 into (3.1.4) we obtain

$$R_Q = \min_{f \in F} R_Q(f) = \int_\Sigma R(f^0, S) dQ(S) = I_Q(\Sigma, A),$$

Q.E.D.

Claim 3.1.1. (B. Ryabko (1979). Channel – universal code relation)
Let Σ be a set of sources. The capacity C of the corresponding channel equals the redundancy of the optimal for Σ encoding to within the additive one:

$$C \le R(\Sigma) \le C + 1.$$

Proof.
Let the cardinality $|A|$ of the alphabet A equal k, I_k be a cube:

$$I_k = \{f : f = (f_1, \dots, f_k), \ 0 \le f_i \le k, \ i = 1, \dots, k\}.$$

When encoding the sources $S \in \Sigma$, we will use only vectors f, for which the Kraft inequality (3.1.1) turns to an equality. Those vectors belong to I_k. Thus,

$$\min_{f \in F} R(f, S) = \min_{f \in F \cap I_k} R(f, S).$$

Consider a zero-sum game of two players, see Blackwell, Girshick (1954). Strategies of the first player are sources $S \in \Sigma$. Strategies of the second one are vectors from $F \cap I_k$.
  If the first player chooses a source S, the second one chooses a vector f, then the pay is $R(f, S)$. This function is continuous and convex with respect to f. The set of strategies $F \cap I_k$ is convex, closed, bounded. There is a Theorem in Blackwell and Girshick (1954), which says that there are both a price of the game and an optimal pure strategy of the second player in that case. It means that

$$\sup_Q \inf_{f \in F \cap I_k} \int_\Sigma R(f, S) dQ(S) = \inf_{f \in F \cap I_k} \sup_Q \int_\Sigma R(f, S) dQ(S) \qquad (3.1.5)$$

Q is a probability measure on Σ. Hence,

$$\sup_Q \int_\Sigma R(f,S)dQ(S) = \sup_{S \in \Sigma} R(f,S) \qquad (3.1.6)$$

We obtain from (3.1.3) and the definition of $R(\Sigma)$:

$$\inf_{f \in F} \sup_{S \in \Sigma} R(f,S) \le R(\Sigma) = \inf_{f \in \lceil F \rceil} \sup_{S \in \Sigma} R(f,S) \le \inf_{f \in F} \sup_{S \in \Sigma} R(f,S) + 1 \qquad (3.1.7)$$

The definition of the capacity C alongside with (3.1.6) and (3.1.5) yield

$$C = \inf_{f \in F} \sup_{S \in \Sigma} R(f,S)$$

This equality and (3.1.7) give the Claim. Q.E.D.

Thus, to find the redundancy of an optimal code we can find the capacity of a channel. There is an algorithm of Blahut to calculate the capacity. It can be applied to obtain the redundancy for a set of sources. In the following Sections we will get explicit expressions for the optimal redundancy in some interesting cases.

Concluding the Section we cite Theorem 4.5 from Gallager (1968). It says that a measure Q on a set Σ gives $\sup_Q I_Q(\Sigma, A)$ iff $I_Q(S, A)$ takes one and the same value on all sources, whose measure is not zero. That Theorem will be used in the next Sections.

3.2 Block-to Variable Length Encoding of Bernoulli Sources

The results of the previous Section are applied next to the Bernoulli sources. A source S is identified with a vector $S = (S_1, \ldots, S_{|B|})$, where S_i is the probability of a letter S_i, $1 \le i \le |B|$, B is an alphabet. The alphabet A of S is the n-th degree of B, $n > 0$, $B^n = A$. A source S produces n-length words in B, which can be considered as letters of A,

$$|A| = k = |B|^n.$$

If a is a letter of A, $r_j(a)$ is the number of occurrences of $j \in B$ in a, then a source S generates a word a with the probability $p_S(a)$,

$$p_S(a) = S^{r(a)}, \qquad (3.2.1)$$

$r(a) = \left(r_1(a), \ldots, r_{|B|}(a) \right).$

Two equations:

$$\sum_{a,\ |a|=n} p_S(a)r(a) = nS \qquad (3.2.2)$$

and

$$\sum_{a,\ |a|=n} p_S(a) r_j^2(a) = n^2 S_j^2 + S_j(1 - S_j)n, \quad j = 1, \ldots, |B| \tag{3.2.3}$$

are easy to verify. They can be found in Feller (1957).
    The letter Σ stands for the set of $|B|$-dimensional vectors

$$S = (S_1, \ldots, S_{|B|}), \quad S_1 + \ldots + S_{|B|} = 1, \quad S_i \geq 0,$$

i.e., Σ is the $|B|$-dimensional simplex. Σ represent the set of Bernoulli sources with $|B|$ letters. The redundancy $R(\Sigma)$ is introduced in 3.1. We are interested in its asymptotic behaviour.
    The empirical entropy

$$H_B(a) = -\sum \frac{r_i(a)}{n} \log \frac{r_i(a)}{n}, \quad |a| = n,$$

has got an important role to play. It was introduced in 1.3.

Lemma 3.2.1. (average empirical entropy)
    Let $p_S(a)$ be the probability for a Bernoulli source S to produce a word a, $|a| = n$, $n > 0$, $H_B(a)$ be the empirical entropy of a, $H(S)$ be the entropy of a source S, Σ be the set of all Bernoulli sources. Then the average of $H_B(a)$ over Σ equals $H(S)$ to within $O\left(\frac{1}{n}\right)$:

$$-\frac{|B| - 1}{n} \log e \leq \sum_{S \in \Sigma} p_S(a) H_B(a) - H(S) \leq 0$$

Proof.
Equality (3.2.1) and the definition of empirical entropy yield:

$$\sum_{a,\ |a|=n} p_S(a) H_B(a) - H(S) = -\sum_a \sum_{j=1}^{|B|} p_S(a) \frac{r_j(a)}{n} \log \frac{r_j(a)}{n S_j} \tag{3.2.4}$$

Apply the Iensen inequality for $-\log x$ to (3.2.4):

$$\sum_{a,\ |a|=n} p_S(a) H_B(a) - H(S) \geq -\log \sum_{j=1}^{|B|} \sum_{a,\ |a|=n} p_S(a) \frac{r_j^2(a)}{n^2 S_j} \tag{3.2.5}$$

Use in (3.2.5), first, equality (3.2.3) and, second, the inequality $\ln x \leq x - 1$. We obtain:

$$\sum p_S(a) H_B(a) - H(S) \geq \frac{-|B| - 1}{n} \log e. \tag{3.2.6}$$

On the other hand, use the Iensen inequality for $-x \log x$. It gives

$$\sum_{a,\ |a|=n} p_S(a) H_B(a) - H(S) \leq 0. \tag{3.2.7}$$

Inequalities (3.2.6) and (3.2.7) give the Lemma. Q.E.D.

Introduce some notations. The symbol $D(S/\lambda)$ stands for the Dirichlet density distribution on the B-dimensional simplex with a parameter $\lambda = (\lambda_1, \ldots, \lambda_{|B|})$:

$$D(S/\lambda) = \frac{\Gamma(|\lambda|)}{\prod\limits_{j=1}^{|B|} \Gamma(\lambda_j)} \prod_{j=1}^{|B|} S_j^{\lambda_j - 1} \tag{3.2.8}$$

The parameter $1/2 = (1/2, \ldots, 1/2)$ happens to be especially important. By E_λ we denote the mathematical expectation with respect to $D(S/\lambda)$:

$$E_\lambda f = \int\limits_{S,\ |S|=1} f(S) D(S/\lambda) dS \tag{3.2.9}$$

For a word a, $|a| = n$, we can easily get

$$E_{1/2} p_S(a) = \Gamma\left(\frac{|B|}{2}\right) \frac{\prod\limits_{j=1}^{|B|} \Gamma(r_j(a) + 1/2)}{\pi^{\frac{|B|}{2}} \Gamma(n + \frac{|B|}{2})} \tag{3.2.10}$$

For a word a and a word c we have

$$E_{1/2} p_S(a) p_S(c) = E_{1/2} p_S(c) E_{r(c)+1/2} p_S(a) \tag{3.2.11}$$

We will next develop an asymptotically optimal code for the set of Bernoulli sources.

A code optimal for Bernoulli sources is a modification of EE – code of Section 2.9. We call it MEE – code. For a word a, $|a| = n$, it consists of two parts. The first one is the prefix. Its length is

$$\frac{|B| - 1}{2} \log n + O(B).$$

The length of the second one is $n H_B(a)$, which coincides with the main addend of EE – code.

Theorem 3.2.1. (universal encoding of Bernoulli sources).

Let Σ be the set of Bernoulli sources over an alphabet B, n be the wordlength, $n \to \infty$. The code MEE, whose length on a word a equals

$$|MEE(a)| = n H_B(a) + \frac{|B| - 1}{2} \log n + O(B)$$

is asymptotically optimal for Σ. The redundancy $R(\Sigma)$ equals asymptotically $\frac{B-1}{2} \log n$.

Proof.

The Iensen inequality and the inequality $\ln(1 + x) \leq x$ yield:

$$0 = \sum_{j=1}^{|B|} \frac{r_j(a)}{n} \log \frac{r_j(a) + 1/2}{n} + H_B(a) \leq \log\left(1 + \frac{|B|}{2n}\right) = O\left(\frac{|B|}{n}\right) \qquad (3.2.12)$$

Use (3.2.12) and the Stirling formula (2.9.11) to rewrite (3.2.10) as

$$-\log E_{1/2}p_S(a) \doteq nH_B(a) + \frac{|B| - 1}{2}\log n + O(B) \qquad (3.2.13)$$

The information between a source S and the output of the corresponding channel was defined in 3.1. Being taken with respect to the Dirichlet measure (3.2.9), it is

$$I_D(S, \Sigma) = \sum_{a,\,|a|=n} p_S(a)\left(\log p_S(a) - \log E_{1/2}p_S(a)\right) \qquad (3.2.14)$$

Substitute into (3.2.14) the equality (3.2.13) and use Lemma (3.2.1). It gives:

$$I_D(S, \Sigma) = \frac{|B| - 1}{2}\log n + O(B).$$

As we can see, the information $I_D(S, \Sigma)$ does not depend on the input S. According to the Theorem cited at the end of Section 3.1, the Dirichlet measure gives the capacity of the channel, which equals $R(\Sigma)$ by Lemma 3.1.1. Q.E.D.

To encode a word a one ought not to know its probability. It is enough to know the frequencies $\frac{r_i(a)}{n}$ only. It does not matter, which source produces words. In any case, per letter redundancy of the code is $O\left(\frac{\log n}{n}\right)$ and goes to zero.

Compare asymptotically optimal code of Theorem 3.1.1 with Empirical Entropic Code of Section 2.9. The main contribution to the codelengths of both codes makes the empirical entropy $|a|\,H(a)$. They differ by their prefix only.

Asymptotically optimal code for the set of Markov sources of any order was constructed by V.K.Trofimov (1974). If u and v are words, $u = u_1, \ldots, u_k$, $v = v_1, \ldots, v_k$, $k = |v| \leq |u| = n$, then $r_v(u)$ is the number of occurrences of v among the words $u_1 \ldots u_k, u_2 \ldots u_{k+1}, \ldots, u_{n-k+1} \ldots u_n$. For Markov sources, whose memory equals k, the optimal is a code, which maps u to a word $MEE^k(u)$,

$$MEE^k(u) = -\log \frac{\Gamma(|B|/2)}{(\Gamma(1/2))^{|B|}\,|B^k|} \prod \frac{\prod\limits_{b \in B} \Gamma(r_{\beta B}(u) + 1/2)}{\Gamma\left(r_\beta(u') + \frac{|B|}{2}\right)}$$

where u' is obtained from u by omitting the last letter. The redundancy of f is $\frac{|B|^k \cdot (|B|-1)}{2} \cdot \log n$.

3.3 Variable-Length-to-Block Encoding of Bernoulli Sources

Such an encoding is defined by a tree Δ. We traverse the tree until a leaf is reached. Then we produce a code at a leaf and make a fresh start. All codes have one and the same length. It is a generalization of Khodak's VB-code for a known Bernoulli source (Section 2.7). Here we are going to develop a VB-code, which is good for all Bernoulli sources simultaneously. We name the code after V. Trofimov, who developed it in Trofimov (1976). We confine ourselves to binary case for simplicity sake: a source S produces 0 or 1 with probabilities S_1 or S_2.

The delay or average height of a coding tree Δ on a source S is given by (2.1.13):

$$d_S(\Delta) = \sum_{a \text{ is a leaf of } \Delta} p(a) |a| \tag{3.3.1}$$

Each leaf of Δ is encoded by a binary word. The lengths of those words are equal to $\lceil \log |L\Delta| \rceil$, where $|L\Delta|$ is the quantity of leaves of Δ. The redundancy of such a BV-code on a source S is $R_S(\Delta)$,

$$R_S(\Delta) = \frac{\lceil \log |L\Delta| \rceil}{d_S(\Delta)} - H(S) \tag{3.3.2}$$

If a sequence $\{\Delta_n\}$ of trees is universal, then

$$\lim_{n \to \infty} R_S(\Delta_n) = 0 \tag{3.3.3}$$

on any source S. We have from (3.3.2) and (3.3.3), that

$$\lceil \log |L\Delta_n| \rceil = d_S(\Delta_n) \cdot H(S)(1 + \mathrm{o}\,(1)) \tag{3.3.4}$$

on any source S.

The probabilities of leaves in Khodak's code are nearly equal to each other. The average probabilities of leaves in Trofimov's code are nearly equal to each other. We will average with respect to the Dirichlet measure with parameter $1/2$, see (3.2.10).

Trofimov's algorithm of universal VB-encoding is as follows. Fix a positive t. Declare a word a to be a leaf of a tree Δ, if, on one hand,

$$E_{1/2} p_S(a) \geq 2^{-t} \tag{3.3.5}$$

On the other hand, either

$$E_{1/2} p_S(a0) < 2^{-t} \tag{3.3.6}$$

or

$$E_{1/2} p_S(a1) < 2^{-t} \tag{3.3.7}$$

As it is seen from (3.3.6), (3.3.7) and (3.2.10),

$$t \leq -\log E_{1/2} p_S(a) + \log(n + 1) + 1 \tag{3.3.8}$$

Start moving from the root until (3.3.5)-(3.3.7) are met.

The corresponding word a will be made a leaf. Start again. We will get a Trofimov's encoding tree Δ. Summing (3.3.5) up over all leaves of Δ, we get

$$|L\Delta| \leq 2^t, \tag{3.3.9}$$

Upperbound $\lceil \log |L\Delta| \rceil$ in (3.3.2) by $t+1$ (due to (3.3.9)). Upperbound t by (3.3.8). Use formula (3.2.13) for $-\log E_{1/2}p_S(a)$. Upperbound the logarithm and entropy by Iensen inequality, employing Lemma 2.1.1:

$$\sum_{a \text{ is a leaf of } \Delta} p(a)r_i(a) = d_S(\Delta)S_i, \quad i = 1, 2.$$

where $r_i(a)$, $i = 1, 2$ is the number of zeroes (ones) in a. We obtain

$$R_S(\Delta) \leq \frac{3/2 \log d_S(\Delta) + C}{d_S(\Delta)} \tag{3.3.10}$$

Let t go to infinity. We have a sequence of trees. The corresponding BV-codes are universal by (3.3.10). Substitute (3.3.4) into (3.3.10). We obtain

Claim 3.3.1.

Let Δ_n be a sequence of Trofimov's trees. The redundancy of the corresponding BV-code on any Bernoulli source $S = (S_1, S_2)$, $S_1 > 0$, is upperbounded by

$$\frac{3}{2} \frac{\log \log |L\Delta|}{\log |L\Delta|} H(S)(1 + o(1))$$

where $|L\Delta|$ is the number of leaves of the coding tree.

This number characterizes the complexity of encoding.

Trofimov's code is nearly optimal. We are going to lowerbound its redundancy.

We say that a tree Δ_2 follows a tree Δ_1 immediately, if a leaf a of Δ_1 is changed in Δ_2 for two words, $a0$ and $a1$. We say that a tree Δ_l follows Δ_1 and write $\Delta_l > \Delta_1$, if there is a chain of trees $\Delta_l, \ldots, \Delta_1$, in which a tree Δ_i follows Δ_{i-1} immediately, $i = 2, \ldots, l$. The average information between a source S and a tree Δ is

$$I(\Delta, S) = \sum_{a \text{ is a leaf of } \Delta} p_S(a) \log \frac{p_S(a)}{E_1 p_S(a)} \tag{3.3.11}$$

where

$$E_1 p_S(a) = \int_0^1 S_1^{r_1(a)} (1 - S_1)^{r_2(a)} dS_1 = \frac{r_1(a)! r_2(a)!}{(|a| + 1)!} \tag{3.3.12}$$

The Stirling formula yields

$$-E_1 p_S(a) = |a|\, H_B(a) + \frac{1}{2}\ln|a| - \frac{1}{2}\ln\frac{r_1(a)}{|a|} - \frac{1}{2}\ln\frac{r_2(a)}{|a|} + O(1) \geq$$

$$\geq |a|\, H_B(a) + \frac{1}{2}\ln|a| + \mathrm{O}\,(1), \qquad\qquad (3.3.13)$$

where $B = \{0,1\}$ is binary alphabet.

Lemma 3.3.1.
   Average information is an increasing function: if $\Delta_2 > \Delta_1$, then $I(\Delta_2, S) \geq I(\Delta_1, S)$.

Proof.
We can assume that Δ_2 follows Δ_1 immediately by changing a word a for $a0$ and $a1$. Then, by (3.3.11) and (3.3.12),

$$I(\Delta_2, S) = I(\Delta_1, S) + p_S(a)\left(S_1 \log\frac{|a|+2}{r_1(a)+1} + S_2 \log\frac{|a|+2}{r_2(a)+1} - H(S)\right) \quad (3.3.14)$$

The numbers
$$l_1 = \log\frac{r_1(a)+1}{|a|+2}, \quad l_2 = \log\frac{r_2(a)+1}{|a|+2}$$

meet the Kraft inequality:
$$2^{l_1} + 2^{l_2} = 1.$$

Thus, by Claim 2.5.1, the bracketed expression in (3.3.14) is nonnegative, Q.E.D.

Claim 3.3.2.
   Let Δ_n, $n = 1, 2, \ldots$ be a sequence of trees, such that $R_S(\Delta_n) \to 0$ on any Bernoulli source S. Then

$$\sup_S R_S(\Delta_n) \geq C \cdot \frac{\log\log|L\Delta_n|}{\log|L\Delta_n|},$$

C be a constant.
   It means that Trofimov's code is optimal to within a multiplicative constant.

Proof.
Let Δ_n be a sequence of binary trees,

$$\gamma_n = \left\lceil \frac{\log|L\Delta_n|}{\log\log|L\Delta_n|} \right\rceil \qquad\qquad (3.3.15)$$

If a is a leaf of Δ_n and the height $|a|$ of a is less than γ_n, then supplement Δ_n by a complete subtree, whose root is a and whose height is $\gamma_n - |a|$. Not more than

$$2^{\gamma_n} = o(\log|L\Delta_n|)$$

nodes would be added. The number of leaves of the tree obtained $\tilde{\Delta}_n$ is less than $2\,|L\Delta_n|$, its delay is increased. Hence, by (3.3.2) and the inequality $d_S(\tilde{\Delta}_n) \geq d_S(\Delta_n)$,

$$R_S(\tilde{\Delta}_n) \leq R_S(\Delta_n) + \frac{1}{d_S(\Delta_n)} \tag{3.3.16}$$

Choose α and β, $0 < \alpha < \frac{1}{2} < \beta < 1$. Obviously,

$$\sup_S R_S(\tilde{\Delta}_n) \geq \frac{1}{\beta - \alpha} \int_\alpha^\beta R_S(\tilde{\Delta}_n) dS, \quad dS = dS_1. \tag{3.3.17}$$

Relation (3.3.4) yields

$$d_S(\tilde{\Delta}_n) \leq \frac{\log \left|L\tilde{\Delta}_n\right|}{\min\left(H(\alpha), H(\beta)\right)}(1 + \mathrm{o}\,(1)),$$

or

$$d_S(\Delta_n) < C \log \left|L\tilde{\Delta}_n\right|, \quad C = \mathrm{const}, \quad n \to \infty. \tag{3.3.18}$$

Rewrite (3.3.17), employing (3.3.18) and Lemma 2.1.1:

$$\sup_S R_S(\tilde{\Delta}_n) \geq$$

$$\geq \frac{C}{\log \left|L\tilde{\Delta}_n\right|(\beta - \alpha)} \int_\alpha^\beta \left(\lceil \log |L\Delta_n| \rceil + \sum_{a \text{ is a leaf of } \Delta_n} p_S(a) \log p_S(a) \right) dS$$

Leaves are given codes of length $\lceil \log \left|L\tilde{\Delta}_n\right| \rceil$. Give a leaf a a code $\lambda(a)$, whose length equals minus logarithm of the average probability of a :

$$|\lambda(a)| = -\log \frac{1}{\beta - \alpha} \int_\alpha^\beta p_S(a) dS \tag{3.3.20}$$

It can only decrease the right side of (3.3.19).

From (3.3.20) we get

$$|\lambda(a)| \geq -\log \frac{1}{\beta - \alpha} \int_0^1 p_S(a) dS = E_1 p_S(a) + C. \tag{3.3.21}$$

Rewrite (3.3.19) as

$$\sup_S R_S(\tilde{\Delta}_n) \geq \frac{C}{\log \left|L\tilde{\Delta}_n\right|} \int_\alpha^\beta \sum_{a \text{ is a leaf of } \tilde{\Delta}_n} p_S(a) \log \frac{p_S(a)}{E_1 p_S(a)} + O\left(\frac{1}{\log \left|L\tilde{\Delta}_n\right|}\right). \tag{3.3.22}$$

We can recognize in (3.3.22) the information $I(\Delta, S)$:

$$\sup_S R_S(\tilde{\Delta}_n) \geq \frac{C}{\log\left|L\tilde{\Delta}_n\right|} \int_\alpha^\beta I(\tilde{\Delta}_n, S)dS + O\left(\frac{1}{\log\left|\tilde{\Delta}_n\right|}\right). \tag{3.3.23}$$

The tree $\tilde{\Delta}_n$ follows the complete uniform tree Δ'_n, whose all leaves have height γ_n :

$$\tilde{\Delta}_n \geq \tilde{\Delta}'_n.$$

By Lemma 3.3.1,
$$I(\tilde{\Delta}_n, S) \geq I(\Delta'_n) \tag{3.3.24}$$

Substitute inequality (3.3.13) into definition (3.3.11) of the average information for the tree Δ'_n. Apply then Lemma 3.2.1. We set

$$I(\Delta'_n, S) \geq \frac{1}{2}\ln \gamma_n + O(1) \tag{3.3.25}$$

Substitute inequalities (3.3.24) and (3.3.25) into (3.3.23). The value of γ_n is given by (3.3.15). It yields

$$\sup R_S(\tilde{\Delta}_n) \geq \frac{C}{\log\left|L\tilde{\Delta}_n\right|} \cdot \log\log|L\Delta_n|(1 + o(1)) \tag{3.3.26}$$

Combine (3.3.26), (3.3.16) and the inequality

$$\left|L\tilde{\Delta}_n\right| \leq 2|L\Delta_n|.$$

We obtain the lower bound sought for. Q.E.D.

3.4 Adaptive Encoding

When making a universal code, we know only, that a source S belongs to a set Σ. When making the Shannon code, we know a source S exactly. There are intermediate situations: we have gathered some information about a source, but that information is not complete.

Suppose we have observed a source S for a time t. As a result, we have got a word w, $|w| = t$, which was produced by S. The word w may be called a sample. Given w, we want to encode S. For each sample w there is a coding map f_w. Hence, we have a family of coding maps

$$\varphi = \{f_w\}, \quad |w| = t.$$

Let a Bernoulli source $S = \{S_1, \ldots, S_{|B|}\}$ emit n-length words in an alphabet B. The redundancy of a map f_w on a source S is

$$R(f_w, S) = \sum_{a, |a|=n} p_S(a)|f_w(a)| - H(S) \tag{3.4.1}$$

The probability $p_S(a)$ is given by (3.2.1)

The redundancy $R(\varphi, S)$ of a family φ on a source S is obtained by averaging $R(f_w, S)$ over all samples w :

$$R(\varphi, S) = \sum_{w, \, |w| = t} p_S(w) R(f_w, S) \qquad (3.4.2)$$

The redundancy $R(\varphi, \Sigma)$ of a family φ on the set Σ of Bernoulli sources is defined by the equation

$$R(\varphi, \Sigma) = \sup_{S \in \Sigma} R(\varphi, S) \qquad (3.4.3)$$

The infimum of $R(\varphi, \Sigma)$ over all families is denoted by $R_t(\Sigma)$:

$$R_t(\Sigma) = \inf_{\varphi} R(\varphi, \Sigma).$$

Our goal is to develop a family φ, for which $R(\varphi, \Sigma)$ equals asymptotically $R_t(\Sigma)$. Such a family is adapting to a source S, when t goes to infinity. If there is no any sample, i.e., $t = 0$, then we have the universal code of the Section 3.2.

There are two kinds of coding families. The first of them is called direct. It gives a code to a n-length word a by a sample w. The second one is indirect. It begins with an estimation of letter probabilities by a sample w. Given this estimation, a word a is encoded.

Theorem 3.4.1. (Adaptive encoding of Bernoulli sources)

Let Σ be the set of Bernoulli sources over an alphabet B, n be the wordlength, t be the sample length, $t \to \infty$, $R_t(\Sigma)$ be the redundancy of the best direct adaptive code, which uses t-length samples.

Then

$$R_t(\Sigma) = \frac{|B| - 1}{2} \log \frac{t + n}{\hat{t}} + R',$$

where $\hat{t} = \max(t, 1)$, $C_1 > R' > C_2$, C_1, $C_2 = \text{const}$.
given a sample w, $|w| = t$, an asymptotically best code f_w takes a word a, $|a| = n$, to its code $f_w(a)$,

$$|f_w(a)| = \lceil -\log E_{r(w) + 1/2} p_S(a) \rceil$$

Proof.
Upper bound.

Upperbound $R(\varphi, \Sigma)$, where $\varphi = \{f_w\}$, f_w is given in the Theorem, $E_{r(w)+1/2} p_S(a)$ is given by (3.2.11). Obviously, the Kraft inequality holds for that code. Rewrite $|f_w(a)|$, using (3.2.11), (3.2.13) and the relation

$$p_S(a) p_S(w) = p_S(aw).$$

It yields

$$|f_w(a)| = \lceil -\log E_{r(w)+1/2} p_S(a) \rceil =$$

$$= (n+t)H(aw) - tH(w) + \frac{|B|-1}{2}\log\frac{n+t}{\hat{t}} + O(B) \tag{3.4.4}$$

Lemma 3.2.1 yields two inequalities for the average values of $H(aw)$ and $H(w)$. Those are

$$\sum_{|a|=n,\ |w|=t} p_S(a)p_S(w)H_B(aw) = H(S) + O\left(\frac{1}{n+t}\right) \tag{3.4.5}$$

and

$$\sum_{|a|=n,\ |w|=t} p_S(a)p_S(w)H_B(w) = H(S) + O\left(\frac{1}{t}\right) \tag{3.4.6}$$

Now we take definitions (3.4.1)-(3.4.3) and substitute expressions (3.4.4)-(3.4.6) into them:

$$R_t(\Sigma) \le \frac{|B|-1}{2}\log\frac{n+t}{\hat{t}} + C,$$

Q.E.D.

Lower bound.

The redundancy $R(\varphi, \Sigma)$ is defined as a supremum. The supremum of a function is not less than its average value with respect to a measure. Choose the Dirichlet distribution $D(S/(1/2))$ as that measure:

$$R_t(\Sigma) = \inf_{\varphi}\sup_{S\in\Sigma} R(\varphi, S) \ge \inf_{\varphi} E_{1/2}R(\varphi, S) \tag{3.4.7}$$

Use in (3.4.7) definitions (3.4.1)-(3.4.3) and formula (3.2.11):

$$R_t(\Sigma) \ge \inf_{\varphi}\sum_{|w|=t} E_{1/2}p_S(w)\sum_{|a|=n} E_{r(w)+1/2}p_S(a)\,|f_w(a)| - E_{1/2}H(S) \tag{3.4.8}$$

The words $f_w(a)$ make a prefix code. The inner sum $\sum_{|a|=n}$ will be minimal, if $|f_w(a)|$ equals the minus log coefficient, with which $|f_w(a)|$ enters that sum. In other words, $\sum_{|a|=n}$ is minimal, if

$$|f_w(a)| = -\log E_{r(w)+1/2}p_S(a). \tag{3.4.9}$$

Use in (3.4.8), first, expression (3.4.4) and then expressions (3.4.5) and (3.4.6):

$$R_t(\Sigma) \ge E_{1/2}\left(\frac{|B|-1}{2}\log\frac{n+t}{\hat{t}} + O(1)\right) \ge \frac{|B|-1}{2}\log\frac{n+t}{\hat{t}} + C \tag{3.4.10}$$

As it follows from definitions, $R(\varphi, \Sigma)$ is not less than the redundancy of the encoding of any given source S'. Let S' be a Bernoulli source with nonconcerted probabilities

(Section 2.7). Since S' is a known source, we can take $n = $ const, $t \to \infty$. Consequently,

$$\log \frac{n+t}{\hat{t}} \to 0,$$

and we have from (3.4.10):

$$R_t(\Sigma) \geq C.$$

On the other hand, the per block redundancy of any code on S' is not less than a positive constant (Theorem 2.7.1). Thus, the constant in (3.4.10) is positive, Q.E.D.

If we let $t = 0$, we will get the case of universal encoding (Section 3.2). Then $R_0(\Sigma) = R(\Sigma)$, and we will once again obtain Theorem 3.2.1 as a corollary of the Theorem 3.4.1.

Suppose one wishes to design an n-length block code for a source S. For how long should one observe S to make a good code? If the observation time t equals the blocklength n, then, by Theorem 3.4.1, the redundancy

$$R_t(\Sigma) = O(1).$$

If t goes to infinity, $R_t(\Sigma)$ remains lowerbounded by a positive constant. It means that there is no much sense in great observation time. That time should be of the same order of magnitude as the blocklength.

Universal encoding is an extreme of the adaptive one ($t = 0$). The other extreme is per letter encoding ($n = 1$). Given a word w, we estimate the probability p_i of a letter $i \in B$ as

$$p(i) = \frac{r_i(w) + 1/2}{t + \frac{|B|}{2}} \tag{3.4.11}$$

It follows from Theorem 3.3.1 and formulas (3.2.10), (3.2.1).

The code $f_w(i)$ of the Theorem 3.4.1 is the Shannon code for the distribution defined by (3.4.11).

The Dirichlet distribution with $\lambda = 1/2$ is the best choice asymptotically, $n \to \infty$. If $n = 1$, then $\lambda = 1$ is acceptable as well. It gives the following estimation of the probability $p(i)$:

$$p(i) = \frac{r_i(w) + 1}{t + |B|} \tag{3.4.12}$$

Either estimation (3.4.11) or (3.4.12) is a starting point for indirect adaptive encoding. A word a, $|a| = n$, is given an indirect code $f^{\text{ind}}(a)$, whose length is

$$\left| f^{\text{ind}}(a) \right| = \sum_{i \in B} r_i(a) \lceil \log p(i) \rceil \tag{3.4.13}$$

The redundancy $R(\varphi^{\text{ind}}, \Sigma)$ of such a family on the set Σ of Bernoulli sources is defined by (3.4.1)-(3.4.3).

Claim 3.4.1. (indirect adaptive code)

Let Σ be the set of Bernoulli sources over an alphabet B, n be the wordlength, t be the sample length, $t \to \infty$, $R(\varphi^{ind}, \Sigma)$ be the redundancy of the indirect adaptive code, determined by (3.4.12). Then

$$R(\varphi^{ind}, \Sigma) \le n \log e \cdot \frac{|B| - 1}{t + 1} + 1.$$

Proof.

Let S be a Bernoulli source, $p_S(i)$ be the probability of a letter i, $i \in B$, the coding family φ^{ind} be defined by (3.4.12) and (3.4.13). We have for the redundancy $R(\varphi^{ind}, \Sigma)$ of φ^{ind} on S :

$$R(\varphi^{ind}, \Sigma) = \sum_{w,|w|=t \; a,|a|=n \; i,i\in B} p_S(w) p_S(a) \lceil \log \frac{(t + |B|) p_S(i)}{r_i(w) + 1} \rceil r_i(a) \qquad (3.4.14)$$

Apply to (3.4.14) the inequality

$$\ln(1 + x) \le x$$

and the equality

$$\sum_{|a|=n} r_i(u) p_S(u) = n p_S(i).$$

We obtain:

$$R(\varphi^{ind}, \Sigma) \le$$

$$\le n \log e \left[\sum_{i \in B} p_S(i) \sum_{w,|w|=t} \frac{(t + |B|) p_S(i)}{r_i(w) + 1} p_S(w) - 1 \right] + n \qquad (3.4.15)$$

Since S is a Bernoulli source, (3.4.15) can be rewritten as

$$R(\varphi^{ind}, \Sigma) \le$$

$$\le n \log e \left[\sum_{i \in B} (p_S(i))^2 \sum_{k=0}^{t} C_t^k (p_S(i))^k (1 - p_S(i))^{t-k} \cdot \frac{t + |B|}{k + 1} - 1 \right] + n \qquad (3.4.16)$$

Denote

$$\sum_{k=0}^{t} C_t^k x^k q^{t-k} \cdot \frac{1}{k + 1}$$

by $g(x)$. Then

$$(x g(x))' = (x + q)^t, \quad g(x) = \frac{1}{x}(x + q)^{t+1} \cdot \frac{1}{t + 1} \qquad (3.4.17)$$

Apply (3.4.17) to (3.4.16) with $x = p_S(i)$, $q = 1 - p_S(i)$. It yields:

$$R(\varphi^{\text{ind}}, \Sigma) \leq n \log e \cdot \frac{|B| - 1}{t + 1} + 1,$$

Q.E.D.

The per letter redundancy is

$$\log e \cdot \frac{|B| - 1}{t + 1} + O\left(\frac{1}{n}\right)$$

An interesting instance of adaptive code is $n = 1$, $|B| \to \infty$, $t \to \infty$. That is, both the alphabet size and the observation time are sufficiently large. The blocklength equals one. We call that case "sliding window" adaptive encoding. A t-long window slides along a word. At each move of the window the entering letter is given a code in accordance with estimation (3.4.12).

As it follows from Theorem 3.4.1, we ought not to take t much larger than $|B|$. We let $t = O(B)$. Use the Gilbert-Moore code once again. By Claim 3.4.1, its per letter redundancy will be a constant.

To do the coding conveniently, prepare a table. Assume that $B = \{1, \ldots, |B|\}$, $|B| = 2^b$, $1 < 2 < \ldots < |B|$. Given a position of the window, the table contains the probabilities of pairs $(p(1) + p(2),\ p(3) + p(4), \ldots)$, quadruples $(p(1) + p(2) + p(3) + p(4), \ldots)$, etc. All the probabilities are $\log \frac{1}{t}$ -digit numbers. Hence, the size of the table is $O\left(|w| \log^2 |B|\right)$. By the definition of the Gilbert-Moore code, to encode $i \in B$ we shall find the probability of the preceding letters. We sum the probabilities of not more than one letter, one pair, etc, all in all $O\left(\log^2 |B|\right)$ bitoperations. Then we add $\frac{1}{2} p(u)$ and take $\lceil -\log p(i) \rceil + 1$ digits of the results. The number of operations remains the same asymptotically.

Slide the window to the right. A letter leaves it, a letter enters. All estimations of the probabilities, except for those letters, remain the same. Not more than two pairs, quadruples, etc are changed. So, to adjust the table to a new position of the window, we have to make not more than $2 \log |w|$ alterations. We come to

Claim 3.4.2. (B.Ryabko, Sliding window adaptive code, 1990)
Sliding window adaptive encoding with the size of the alphabet equal to $|B|$ and the size of the window equal to $O(|B|)$ has the per letter redundancy $O(1)$, space consumption $O\left(|B| \log |B|\right)$ bits, time consumption $O(\log^2 |B|)$ bitoperations.

3.5 Monotone Sources

Let $A = \{A_1, \ldots A_k\}$, $k > 1$, be an alphabet, S be a stochastic source, $p_S(A_i)$ be the probability to produce A_i by S, $i = 1, \ldots, k$. A source S is identified with the vector $(p_S(A_1), \ldots, p_S(A_k))$. We say that Σ is a set of monotone sources, if for every

$S \in \Sigma$ $p_S(A_1) \geq \ldots \geq p_S(A_k)$. In other words, a source is monotone, if its letters are ordered according to their probabilities. Those probabilities are unknown. The only thing known is their ordering. Nevertheless, it will suffice for making a good enough code, which is not much worse, than one designed for a particular source.

Let t^i be a k-dimensional vector, whose i first coordinates equal $1/i$, and the last $k - i$ coordinates equal zero:

$$t^1 = (1, 0, \ldots, 0), \quad t^2 = (1/2, 1/2, \ldots, 0), \ldots.$$

Lemma 3.5.1.

The set of monotone sources over a k-letter alphabet, $k > 1$, is a convex hull of the vectors $t^1, \ldots t^k$.

Proof.

Let $S = (S_1, \ldots, S_k)$ be a vector from the set Σ,

$$\gamma_i = i(S_i - S_{i+1}), \quad i = 1, \ldots, k, \quad \gamma_{n+1} = 0.$$

By the definition of the set Σ, all numbers $\gamma_1, \ldots, \gamma_k$ are nonnegative. It is easy to verify that, first,

$$\sum_{i=1}^{k} \gamma_i = \sum_{i=1}^{k} S_i = 1,$$

and, second,

$$\sum_{i=1}^{k} \gamma_i t^i = S.$$

It means that Σ is a convex null of $t^1, \ldots, t^k$. Q.E.D.

Theorem 3.5.1. (B. Ryabko, Universal monotone code).

Let Σ be the set of monotone sources over a k-letter alphabet, $R(\Sigma)$ be the redundancy of universal encoding of Σ, see Section 3.1. Then

$$R(\Sigma) = \log \sum_{i=1}^{k} \frac{1}{i} \left(1 - \frac{1}{i}\right)^{i-1} + \alpha_k, \quad |\alpha_k| \leq 1.$$

The code

$$A_i \rightarrow f(A_i), \quad |f(A_i)| = \left\lceil -\log \left(\frac{1}{i}\left(1 - \frac{1}{i}\right)^{i-1} \cdot 2^{-R(\Sigma)}\right)\right\rceil$$

Is optimal to within an additive unit.

If $k \rightarrow \infty$, then

$$R(\Sigma) \sim \log \log k,$$

$$|f(A_i)| \leq \log i + \log \log k + O(1).$$

Proof.
We calculate $R(\Sigma)$ by Claim 3.1.1. Take a channel, whose inputs are sources $S \in \Sigma$ and outputs are letters of the alphabet. The set Σ is a polyhedron:

$$\Sigma = \{S = (S_1, \ldots, S_k), \ S_1 \geq S_2 \geq \ldots S_k \geq 0, \ S_1 + \ldots + S_k = 1\}$$

The redundancy $R(f, S)$ of a code f on a source S is convex with respect to S :

$$R(f, S) = \sum_{i=1}^{k} S_i (\log S_i + f_i)$$

Thus, $R(f, S)$ attains its maximal value at a vertex of the polyhedron Σ. We can omit all inputs of the channel but those vertices, which are $t^1, \ldots, t^k$. So, consider a channel, whose inputs are $t^1, \ldots, t^k$. Let $Q = (q_1, \ldots, q_k)$ be a probability distribution on the input vectors. The probability to get a letter A_i, $i = 1, \ldots, k$, as an output is

$$p_i = \sum_{j=1}^{k} q_j t_i^j, \quad t^j = (t_1^j, \ldots, t_k^j). \tag{3.5.1}$$

Obviously,

$$\sum_{i=1}^{k} p_i = 1 \tag{3.5.2}$$

A Theorem of Gallager was cited at the end of Section 3.1. By that theorem, we have to find a constant C and a distribution Q such that the information between each input vector and the output is C :

$$I_Q(t^j, A) = \sum_{i=1}^{k} t_i^j (\log t_i^j - \log p_i) = C \tag{3.5.3}$$

We solve the system (3.5.3) easily and express p through C. Then from (3.5.1) we find Q. The normalizing equation (3.5.2) gives the constant C, which is the capacity of the channel. As a result, we obtain

$$C = \log \sum_{i=1}^{k} \frac{1}{i} \left(1 - \frac{1}{i}\right)^{i-1}, \quad p_i = 2^{-C}(i-1)^{i-1} \cdot i^{-i},$$

$$q_i = (p_i - p_{i+1})i, \quad p_{k+1} = 0, \quad i = 1, \ldots, k.$$

the functions $\frac{1}{i}$ and $\left(1 + \frac{1}{i-1}\right)^{1-i}$ being decreasing, the numbers q_i are positive. Thus, the distribution Q gives the capacity of the channel defined. The logarithm of p_i is the length of an optimal code of A_i. Q.E.D.

Optimal prefix codes for monotone sources with not more than 9 letters are on Fig. 3.5.1.

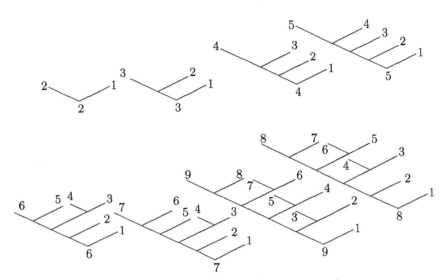

Fig. 3.5.1. Universal monotone code.

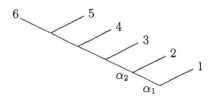

Fig. 3.5.2. A key for Libellulinae dragon-flies.

Those codes were used by Krichevskii, Ryabko, Haritonov (1981) to make identifying keys.

Identifying keys are dichotomous. They are represented by binary trees. At every vertex of a tree we make a decision. If an object to be identified has got a symptom, we go to the right. If it has not, go to the left. There are names of objects at leaves.

An identifying key for dragon-flies of Subfamily Libellulinae is on Fig. 3.5.2. It is taken from Belichev (1963).

That subfamily consists of six genera: 1. Lencorrhinia, 2. Libellula, 3. Pantala, 4. Orthetrum, 5. Neurotemis, 6, Sympetrum. The attributes used to identify are of no interest here.

The frequencies of those genera are not known. However, it is known, that the rarest genus in Siberia is Sympetrum (6). After it goes 5, then 4,3,2,1. So, one may consider the dragon-fly genera as letters of a monotone source.

Take the tree with six leaves on Fig 3.5.1. Then it takes 4 attributes to identify genera 5,6,4 and 3;2 attributes to identify 2;1 - to identify 1. It turns out that it is possible to choose attributes so as realize that tree.

This tree makes the identifying job easier.

Both the number of attributes to be examined and the probability of error are diminished.

There is another way to encode monotone sources. The alphabet $A = \{A_1, \ldots, A_k\}$ may be either finite or infinite.

First we will prove an inequality of Wyner (1972).

Lemma 3.5.2.

Let S be a monotone source with an alphabet A, $p(A_1) \geq p(A_2) \geq \ldots$. Then its entropy upperbounds the mathematical expectation of the logarithm of the letter rank:

$$\sum_{i=1}^{\infty} p(A_i) \log i \leq - \sum_{i=1}^{\infty} p(A_i) \log p(A_i) = H(S)$$

Proof.
The source S being monotone, we have for any $i \geq 1$:

$$1 \geq \sum_{l=1}^{i} p(A_l) \geq ip(A_i) \tag{3.5.4}$$

From (3.5.4):

$$\log i \leq \log \frac{1}{p(A_i)} \tag{3.5.5}$$

averaging (3.5.5) gives Wyner' inequality. Q.E.D.

Now let Σ be the set of all monotone sources over a finite or infinite alphabet, $R(f, S)$ be the redundancy of a code f on a source $S \in \Sigma$,

$$\rho'(f, S) = \frac{R(f, S)}{H(S)}$$

be its relative redundancy.

The following Claim was proved by P. Elias (1975).

Claim 3.5.1.

The relative redundancy of the Levenstein code on any monotone source goes to one as the entropy $H(S)$ goes to infinity:

$$\rho'(\text{Lev}, S) \to 1, \quad \text{as } H(S) \to \infty.$$

For the redundancy $R(\text{Lev}, S)$ we have

$$R(\text{Lev}, S) \leq C \cdot \log H(S).$$

Proof.
The Levenstein code was constructed in Section 2.2. It gives the i-th letter A_i the code Lev (i), where

$$|\text{Lev}(i)| \leq \log i + C \cdot \log\log i, \tag{3.5.6}$$

see (2.2.14).

Averaging (3.5.6) with respect to a distribution $p(A_i)$ and the Wynner inequality (Lemma 3.5.2) give

$$\sum p(A_i) |\text{Lev}(i)| \leq H(S) + C \sum_{i=1}^{\infty} p(A_i) \log\log i \tag{3.5.7}$$

Use in (3.5.7) the Iensen inequality for the logarithm and once again the Wyner inequality. It gives

$$R(\text{Lev}, S) \leq C \log \sum_{i=1}^{\infty} p(A_i) \cdot \log i = C \log H(S)$$

Q.E.D.

Both Levenstein code and the code of the Theorem 3.5.1 can be employed to encode monotone sources. The Levenstein code is preferable if the cardinality of the alphabet is not known beforehand.

Next we apply the codes for monotone sources to the "Move to Front" scheme, which is described in Section 2.9. Let A be an alphabet, $i \to f(i)$ be the code of Theorem 3.5.1. The letters of A constitute a pile. If the first letter of a word w is located at the i-th place in the pile, we transmit $f(i)$, move the letter to the front of the pile and go to the second letter.

Let $a \in A$ and let it occupy the positions $t_1, \ldots t_{r_a(w)}$ in a word w. When being encoded for the first time, a occupies the t_1-th position in the word. Its position in the pile is not lower than t_1-th. When being encoded for the i-th time, it occupies the t_i-th position in the word. Its position in the pile is not lower than $t_i - t_{i-1}$, $i = 2, \ldots, r_a(w)$. Suppose that the total length of codes for all positions of a is $|f_a(w)|$. Then

$$|f_a(w)| \leq |f(t_1)| + \sum_{i=2}^{r_a(w)} |f(t_i - t_{i-1})| \tag{3.5.8}$$

The codelength $|f(t)|$ is defined by Theorem 3.5.1. Apply to (3.5.8) the Iensen inequality for the logarithm. It yields

$$|f_a(w) - r_a(w)(\log\log|A| + C)| \leq$$

$$\leq r_a(w)\log\frac{1}{r_a(w)}\left(t_1 + t_2 - t_1 + \ldots t_{r_a(w)} - t_{r_a(w)-1}\right) =$$

$$= r_a(w)\log\frac{t_{r_a(w)}}{r_a(w)} \leq -r_a(w)\log\frac{r_a(w)}{|w|} \qquad (3.5.9)$$

Sum up the inequality (3.5.9) over $a \in A$. Take into account that the codelength of a word w is $\sum_a |f_a(w)|$. Its empirical entropy is

$$H_A(w) = -\sum_a \frac{r_a(w)}{|w|}\log\frac{r_a(w)}{|w|}$$

We get

Claim 3.5.2. (Move to Front Code)
Let w be a word in alphabet A, $H_A(w)$ be its empirical entropy. The length of Move to Front code of w is not greater than

$$|w|\,H_A(w) + (\log\log|A| + C)\,|w|$$

If we change letters of a word for their binary codes, then the best code is Huffman's. It requires the knowledge of the probabilities. The Move to Front code does not require it. It is $\log\log|A| + C$ bits longer (per letter). Huffman's code may be called two pass. First, we pass w collecting frequencies $r_a(w)$, $a \in w$. Second, we construct a code. Move to Front is one pass code. The word is encoded on line, without any preliminary study.

3.6 Universal Encoding of Combinatorial Sources

We will discuss in this Section just the same problem as in 3.1. We are given a set $\Sigma = \{S\}$. This time Σ consists ot combinatorial sources instead of stochastic ones. Each source S is a subset of a set A. The redundancy of a code f on a source S is

$$R(f, S) = \sup_{x \in S} |f(x)| - H(S),$$

if A is a finite set. If A is a metric space, then we define the ε-redundancy:

$$R_\varepsilon(f, S) = \sup_{x \in S} |f(x)| - H_\varepsilon(S).$$

A code f maps an element x to a binary word $f(x)$, through which x can be recovered to within an ε-error. The redundancy of f on Σ is

$$R_\varepsilon(f, \Sigma) = \sup_{S \in \Sigma} R_\varepsilon(f, S)$$

The ε-redundancy of an optimal code is

$$R_\varepsilon(\Sigma) = \inf_f R_\varepsilon(f, \Sigma),$$

infimum is taken over all prefix codes f.

The redundancies $R(f, \Sigma)$ and $R(\Sigma)$ are defined the same way for a finite set A. Let A be a finite set, $x \in A$,

$$v(x) = \min_{x \in S} |S|, \quad \beta = \sum_{a \in A} (v(a))^{-1}, \quad p(x) = (v(x) \cdot \beta)^{-1} \qquad (3.6.1)$$

As it is easily seen,

$$\sum_{x \in A} p(x) = 1.$$

So, $p(x)$ may be considered a probability distribution on A. Make for $p(x)$ the Shannon code (Claim 2.5.3) and call it r-code for Σ,

$$|r(x)| = \lceil \log p(x) \rceil.$$

For a metric space A the construction is slightly modified. Given $\varepsilon > 0$, we let

$$v(x) = \inf_{x \in S} H_\varepsilon(S).$$

A set Φ_i is defined by

$$\Phi_i = \{x : \lfloor v(x) \rfloor = i\}, \quad i = 0, 1, \ldots, \lfloor H_\varepsilon(A) \rfloor$$

On every set Φ_i we take a minimal ε-net. We give the probability

$$p(x) = 2^{-i} \beta^{-1}, \quad \beta = \sum_{i=0}^{\lfloor H_\varepsilon(A) \rfloor} 2^{H_\varepsilon(\Phi_i) - i},$$

to each element x of that set. We obtain a probability distribution on the union of all ε-nets. Make the Shannon code for that distribution and call it r-code for Σ. If g is an element of an ε-net for Φ_i, $i = 1, \ldots, \lfloor H_\varepsilon(A) \rfloor$, then

$$|r(g)| \leq \log \beta + i + 1 \qquad (3.6.2)$$

To find the r-code of an element $x \in A$, we first take Φ_i, $i = v(x)$. After that we take the nearest to x element of the ε-net of Φ_i. The r-code of that element is, by definition, the r-code. of x.

Claim 3.6.1. (B. Ryabko (1980), a universal combinatorial code.)

Let Σ be a set of combinatorial sources, $\Sigma = \{S\}$, $\cup_{S \in \Sigma} S = A$. Then r-code is optimal for Σ to within an additive constant. The optimal redundancy

$$R(\Sigma) = \log \beta \pm 1$$

For a compact A,

$$\log \beta \geq R_\varepsilon(\Sigma) \geq \log \beta - 2.$$

Proof.
We consider the case of a compact infinite set A. The case of finite A is analogous. If $x \in A$, $\lfloor v(x) \rfloor = i$, then the minimal over $S \in \Sigma$ codelength, through which x can be recovered with ε-error, is i. The length of the r-code exceeds it by $\log \beta + 1$. The upper bound is proved.

Now let f be an arbitrary code,

$$\Psi_i = \cup_{j=0}^{i} \Phi_i, \quad i = 0, \ldots, \lfloor H_\varepsilon(A) \rfloor.$$

If w is a code of a $x \in \Psi_i$, then, by the definitions of Φ_i, Ψ_i and $R_\varepsilon(f, \Sigma)$,

$$|w| \leq R_\varepsilon(f, \Sigma) + i + 1 \tag{3.6.3}$$

The code f being prefix, the Kraft inequality holds:

$$\sum_w 2^{-|w|} \leq 1 \tag{3.6.4}$$

The sum is taken over all w, which are codes of elements of A. The number of words, which are codes of some $x \in \Psi_i$, is $2^{H_\varepsilon(\Psi_i)}$. There are not more than $2^{H_\varepsilon(\Psi_i)} - 2^{H_\varepsilon(\Psi_{i-1})}$ words, which are codes of both a word $x \in \Psi_i$ and a word $y \in \Psi_i$. Use that and (3.6.3) to rewrite (3.6.4) as

$$2^{H_\varepsilon(\Psi_0)} \cdot 2^{-R_\varepsilon(f,\Sigma)+1} + \sum_{i=1}^{\lfloor H_\varepsilon(A) \rfloor} 2^{-R_\varepsilon(f,\Sigma)+i+1} \left(2^{H_\varepsilon(\Psi_i)-H_\varepsilon(\Psi_{i-1})} \right) \leq 1. \tag{3.6.5}$$

We deduce from (3.6.5) that

$$R_\varepsilon(f, \Sigma) \geq \frac{1}{4} \sum_{i=1}^{\lfloor H_\varepsilon(A) \rfloor} 2^{H_\varepsilon(\Psi_i)-i} \tag{3.6.6}$$

Obviously,

$$H_\varepsilon(\Psi_i) \geq H_\varepsilon(\Phi_i), \quad i = 0, \ldots, \lfloor H_\varepsilon(A) \rfloor \tag{3.6.7}$$

Inequalities (3.6.6) and (3.6.7) yield the lower bound of the Claim, Q.E.D.

Next we give two examples of making universal codes for sets of combinatorial sources. The first one concerns identifying keys. In Section 3.5 we knew how the taxons were ordered by their probabilities. Now we know even less than that. There are both a set A of taxons and a set $\Sigma = \{S\}$ of subsets of A. We would like to have a key f such that

$$\max_{S \in \Sigma} R(f, S) = \text{minimum}.$$

There is a key to identify ants of Subgenus Serviformica in Belychev (1963). This Subgenus is present in any part of Russia. Russia is subdivided into five climatic zones: tundra, forest, partially wooded steppe, steppe and desert. Every species of the Subgenus inhabits some ot those zone. Say, species Fusca inhabit forest and partially wooded steppe only. Table 3.6.1 shows which zone is inhabited by which species.

Table 3.6.1.

The distribution of Subgenera Seviformica throughout Russia

N	species	tundra	forest	partially wooded steppe	steppe	desert	the quantity of attributes required to identify	
							key Fig.3.6.2	key Fig.3.6.1
1	Fusca	−	+	+	−	−	4	7
2	Lemani	−	+	−	−	−	3	7
3	Picla	−	+	+	+	−	4	4
4	Garates	+	+	−	−	−	3	6
5.	Koslovi	−	+	−	−	−	4	6
6	Cinerea	−	+	+	+	−	4	2
7	Subpilosa	−	−	−	+	+	4	4
8	Cunguinea	−	+	+	+	+	3	5
9	Rubifarbis	−	+	+	+	+	3	5
10	Garates	−	−	+	−	−	4	6
11	Uralensis	−	+	+	+	−	3	1

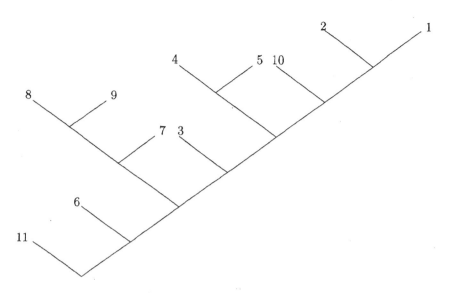

Fig. 3.6.1. Identifying key for Subgenera Serviformica from Belychev (1963).

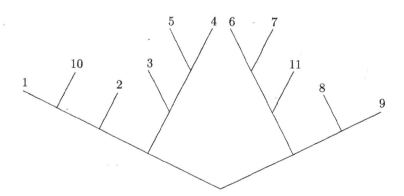

Fig. 3.6.2. An identifying key for Subgenera Serviformica based on r-code.

A new key based on r-code is developed in Krichevskii, Ryabko (1985).

The set A of the Claim 3.6.1 consists now of five subsets: ants of tundra, etc., ants of desert. For a species x find the cardinality $v(x)$ of a minimal subset which contains x. We obtain the following set of numbers: $(7, 9, 6, 1, 9, 6, 3, 3, 3, 7, 6)$. Then we find

$$\beta = \sum_x (v(x))^{-1}$$

and the probabilities

$$p(x) = (v(x) \cdot \beta)^{-1}.$$

Construct the Shannon code for those probabilities and then the corresponding binary tree. One should select attributes for the key according to that tree. Since it can not be implemented exactly, the tree obtained differs slightly from Shannon's one. The tree from the book Belychev (1963) is on Fig. 3.6.1, the key by Claim 3.6.1 is on Fig. 3.6.2. It takes much less time to identify ants by key Fig. 3.6.2.

In second example the set A is a segment $[0, a]$, $a > 0$, the set Σ consists of all subsegments $[0, S]$, $S < a$. Our aim is to develop a universal code for Σ.

Let

$$\varepsilon > 0, \quad \alpha_0 = [0, \varepsilon], \quad \alpha_1 = [\varepsilon, 2\varepsilon], \ldots, \alpha_i = [2^i \varepsilon, 2^{i+1} \varepsilon], \quad i = 0, \ldots \left\lceil \log \frac{a}{2\varepsilon} \right\rceil.$$

A minimal code $r(x)$ of a point x, $x \in A$, consists of two parts. The first one is the number of a segment α_i, $x \in \alpha_i$. The second one is the number of a nearest to x point of an ε-net on α_i. The first part is $\left(\log \log \frac{a}{2\varepsilon} + 1 \right)$ bits long. The second part is $\log \frac{|\alpha_i|}{2\varepsilon} \le i - 1$ bits long. The ε-redundancy does not exceed $\log \log \frac{a}{2\varepsilon} + 2$ bits. Giving to every point of $[0, a]$ the number of a nearest ε-net for $[0, a]$, we obtain the redundancy $\log \frac{a}{\varepsilon} + O(1)$, which is much more than for r-code.

3.7 Universal Encoding of Stationary Sources

We apply the theory of universal encoding to stationary sources. Let $\Sigma = \{S\}$ be a set of such sources over an alphabet A, $F = \{f_1, f_2, \ldots, f_n, \ldots\}$ be a family of prefix codes. A function f_i maps each i-length word to a binary code, $i = 1, \ldots$. The per letter cost of f_i on a source S is defined by (2.1..11):

$$C(f_i, S) = \frac{1}{i} \sum_{|x|=i} |f(x)| \, p(x) \tag{3.7.1}$$

The redundancy of f_i on S is

$$R(f_i, S) = C(f_i, S) - H(S) \tag{3.7.2}$$

A family F is called weakly universal for a set Σ, if

$$\lim_{i \to \infty} R(f_i, S) = 0$$

for every $S \in \Sigma$. If $R(f_i, S)$ converges to 0 uniformly on Σ, i.e., if

$$\lim_{i \to \infty} \sup_{S \in \Sigma} R(f_i, S) = 0,$$

then F is called strongly universal. Given $\varepsilon > 0$ and $S \in \Sigma$, take one by one maps from a weakly universal family. Sooner or later we will come across a good map f_i, whose redundancy on S is less than ε. The number i depends on S. So, if we do not know, which $S \in \Sigma$ we are dealing with, then we will never know that a map taken is good. For a strongly universal family there is one and the same number $i_0 = i_0(\varepsilon)$ which suits all sources $S \in \Sigma$. So, even if we do not know, which S we are dealing with, we will be sure, that a map taken is good, provided $i > i_0$.

Let for a natural i

$$n = n(i) = \left\lceil \log_{|A|} i - 2 \log_{|A|} \log_{|A|} i \right\rceil \tag{3.7.3}$$

Given i, define an alphabet B by

$$B = A^n.$$

The letters of B are n-length words of A.

Take a word $x \in A^i$, $i > 1$, and subdivide it into n-length words. The last word may happen to be shorter. In that case add to it several letters up to the length n. Now x may be considered a word in B, $|x| = \frac{i}{n}$.

We have developed a MEE code f_i for the set of Bernoulli sources in Theorem 3.2.1. A word x in B is mapped to MEE $_i(x)$, which consists of a prefix and a main part:

$$\text{MEE }_i(x) = \frac{i}{n} H_B(x) + \text{prefix}(x),$$

$$|\text{prefix}(x)| = \frac{|B| - 1}{2} \log \frac{i}{n} + O(B),$$

$$H_B(x) = -\sum_{w \in B} \frac{r_w(x)n}{i} \log \frac{i r_w(x)}{n} \tag{3.7.4}$$

A family of coding maps

$$\text{MEE} = \{\text{MEE }_1, \dots, \text{MEE }_i, \dots\}$$

is called Modified Empirical Entropic family.

Theorem 3.7.1. (V. Babkin, Y. Shtarkov [1971])
MEE *-family is weakly universal for the set of all stationary sources. If there is a strongly universal family for a set Σ of stationary sources, then* MEE *-family is also strongly universal for Σ.*

Proof.
1. Let S be a stationary source over an alphabet A and $\varepsilon > 0$. By Claim 2.5.1, there is a number n_0 such that for any $n > n_0$ there exists a prefix map φ_n, which encodes n-length words and has small redundancy:

$$R(\varphi_n, S) \le \varepsilon/2 \tag{3.7.5}$$

Take a natural i such that the number $n(i)$ defined by (3.7.3) is not less than n_0. Encode a word x, $|x| = i$, by φ_n:

$$|\varphi_n(x)| = \sum_{w \in B} r_w(x) \varphi_n(w) = \frac{i}{n} H_B(x) \tag{3.7.6}$$

Here $B = A^n$, $H_B(x)$ is the empirical entropy of x with respect to B. By Claim 2.5.1, the per letter length of any prefix code of x is not less than its empirical entropy:

$$\frac{n}{i} |\varphi_n(x)| \ge H_B(x) \tag{3.7.7}$$

On the other hand, encode x by the MEE $_i$-code. Condition (3.7.3) yields a uniform upperbound for the per letter length of its prefix:

$$\frac{1}{i} |\text{prefix}\,(x)| = \text{O}\left(\frac{1}{\log i}\right) \tag{3.7.8}$$

Bounds (3.7.7) and (3.7.8) give a bound for per letter cost of MEE $_i$ on x:

$$\text{MEE}\,_i(x) \le \frac{1}{i} \varphi_n(x) + \text{O}\left(\frac{1}{\log i}\right) \tag{3.7.9}$$

Averaging (3.7.9) with respect to $p(x)$ upperbounds the per letter costs of MEE $_i(x)$ and $\varphi_n(x)$ on S, i being large enough:

$$C(\text{MEE}\,_i, S) \le C(\varphi_n, S) + \varepsilon/2 \tag{3.7.10}$$

Relations (3.7.2), (3.7.5) and (3.7.10) yield

$$\lim_{i \to \infty} R(\text{MEE}\,_i, S) = 0,$$

i.e., MEE-family is weakly universal.

2. Now let Σ be a set of stationary sources and there be a strongly universal code for Σ. It means that for any $\varepsilon > 0$ there is a code φ_n with uniformly small redundancy:

$$\sup_{S \in \Sigma} R(\varphi_n, S) < \varepsilon/2$$

Take the same code MEE $_i$. Bounds (3.7.7)–(3.7.10) being uniform on Σ, we get

$$\sup_{S \in \Sigma} R(\text{MEE}, S) \leq \varepsilon/2,$$

i.e., MEE-family is strongly universal for Σ, Q.E.D.

Corollary 3.7.1.
There is a strongly universal code for a set Σ iff the per letter entropy goes to its limit uniformly on Σ:

$$\lim_{n\to\infty} \sup_{S\in\Sigma} \left| \frac{1}{n} H(S^1 \vee \ldots \vee S^n) - H(S) \right| = 0.$$

Here S^i is the partition, made by the i-th letter of S, $i = 1, 2, \ldots$.
Proof.
   1. Let φ_n be a code for which

$$\sup_{S\in\Sigma} R(\varphi_n, S) < \varepsilon,$$

i.e.,

$$C(\varphi_n, S) \leq H(S) + \varepsilon \qquad (3.7.11)$$

for any $S \in \Sigma$.
   On the other hand, the per letter cost of a n-block code is not less than per letter entropy, by Claim 2.5.1:

$$C(\varphi_n, S) \geq \frac{1}{n} H(S^1 \vee \ldots \vee S^n). \qquad (3.7.12)$$

By Claim 1.4.2,

$$\frac{1}{n} H(S^1 \vee \ldots \vee S^n) \geq H(S) \qquad (3.7.13)$$

Inequalities (3.7.11)–(3.7.13) yield

$$\sup_{S\in\Sigma} |C(\varphi_n, S) - H(S)| < \varepsilon,$$

The necessity is proven.
   2. Let $\frac{1}{n} H(S^1 \vee \ldots \vee S^n)$ goes to $H(S)$ uniformly, i.e., for any $\varepsilon > 0$ there is $n(\varepsilon)$ such that if $n > n(\varepsilon)$

$$\frac{1}{n} H(S^1 \vee \ldots \vee S^n) - H(S) < \varepsilon/2. \qquad (3.7.14)$$

Take an integer i such that $n(i)$ by (3.7.3) is not less than $n(\varepsilon)$. Denoting by $\text{Sh}_n(S)$ the Shannon code for a source S with codelength n, we have

$$C(\text{Sh}_n(S), S) - \frac{1}{n} H(S^1 \vee \ldots \vee S^n) < \frac{1}{n}. \qquad (3.7.15)$$

Inequality (3.7.7) holds for any encoding of n-length subwords of a word x:

$$C(\mathrm{Sh}_n(S), x) \geq \frac{1}{n} H_B(x). \tag{3.7.16}$$

Relations (3.7.4), (3.7.8), (3.7.14)–(3.7.16) yield

$$\frac{1}{i} \mathrm{MEE}_i(x) = \frac{1}{n} H_B(x) + \mathrm{O}\left(\frac{1}{\log i}\right) \leq$$

$$\leq C(\mathrm{Sh}_n(S), x) + \mathrm{O}\left(\frac{1}{\log i}\right) \leq \frac{1}{n} + \frac{1}{n} H(S^1 \vee \ldots \vee S^n) \leq$$

$$\leq \frac{1}{n} + \frac{\varepsilon}{2} + H(S). \tag{3.7.17}$$

Inequality (3.7.17) shows uniform convergence of $R(\mathrm{MEE}_i, S)$ to zero, i.e., MEE-family is strongly universal for Σ.

Corollary 3.7.2.
Let Σ be a compact set in a metric space, $H(S)$ and $H(S^1 \vee \ldots \vee S^n)$ be continuous functions. Then there is a strongly universal code for Σ.

Proof.
The sequence of continuous functions $\frac{1}{n} H(S^1 \vee \ldots \vee S^n)$ converges to a continuous function $H(S)$ monotonically by Claim 1.4.2. The set Σ being compact, the convergence is uniform by the Dini Theorem. It remains to use Corollary 3.6.1 to obtain the conclusion. Q.E.D.

Define the redundancy $R_n(\Sigma)$ of an optimal code with the blocklength n as

$$R_n(\Sigma) = \inf_{f_n} \sup_{\Sigma} R(f_n, \Sigma),$$

where f_n maps n-length words to their codes. As it was proved in Corollary 3.7.2, $R_n(\Sigma)$ exists for a compact Σ. V. K. Trofimov has shown that

$$R_n(\Sigma) = H_{1/\sqrt{n}}(\Sigma) + \mathrm{O}(1),$$

where $H_\varepsilon(\Sigma)$ is the ε-entropy of Σ.

MEE-family is weakly universal for the set of all stochastic stationary sources. Now we introduce a ECE-family, which plays an analogous role for the set of all combinatorial stationary sources.

The Empirical Combinatorial Entropic map was introduced in Section 2.9. We denote that map by ECE$_i$, if the lengths of words in its domain and the parameter n are related by 3.7.3. We define a ECE-family as

$$\mathrm{ECE} = \{\mathrm{ECE}_1, \ldots, \mathrm{ECE}_i, \ldots\}.$$

As it is explained in Section 2.9, ECE $_i(w)$, $|w| = i$, consists of four parts:
a) word length b) the length of n c) code of V d) main part
The lengths of a) and b) are o$(\log i)$. The length of c) is O $\left(\frac{i}{\log i}\right)$ by its definition
(2.9.7), upperbound (1.1.4) and condition (3.7.3).

The length of c) is $iCH_n(w)(1 + o(1))$, $i \to \infty$. Collecting those estimations, we conclude that

$$|\text{ECE }_i(w)| = iCH_n(w)(1 + o(1)). \tag{3.7.18}$$

Theorem 3.7.2. (universal combinatorial code)
ECE *-family is weakly universal for the set of all stationary combinatorial sources.*
Proof.
We consider a stationary source $S \subseteq A^*$. S consists of finite words. By Claim 1.5.2, there is

$$h(S) = \lim_{l \to \infty} \frac{h(l)}{l},$$

$h(l)$ being the logarithm of the number of l-length words in S. Hence, for any $\varepsilon > 0$ there is n_0 such that

$$h(n) \leq (h(S) + \varepsilon)n, \qquad n > n_0. \tag{3.7.19}$$

S being stationary, the vocabulary of n-length subwords of a word w, $w \in S$, consists of not more than $h(n)$ words. Inequality (3.7.19) and the definition of $CH_n(w)$ yield:

$$CH_n(w) \leq (h(S) + \varepsilon). \tag{3.7.20}$$

Compare (3.7.18) with (3.7.20). We can say, that per letter length of ECE $_i$ code of a word w goes to the combinatorial entropy of a stationary source S, which has generated w, if $|w| = i$ goes to infinity. In other words, ECE-family is weakly universal on the set of all combinatorial stationary sources. Q.E.D.

To compress a long enough word W one has to choose a parameter n and to build a vocabulary of n-length subwords of W. Changing any subword to its dictionary number gives an universal ECE-code.

Including words of different lengths into the vocabulary, we get VB or VV universal codes.

NOTES

To construct the Shannon or Huffman code one needs to know the probabilities of letters, which are never known exactly. It is a drawback of those codes. It came as a surprise that the exact knowledge is not so much needed actually. Boris Fitingof (1966) developed a single code, which is good enough for each Bernoulli source. He campaigned enthusiastically for that code at Dubna (Russia) International Conference on Information Theory. His paper was inspired by A. N. Kolmogorov (1965). It is fair to consider B. Fitingof the founder of the universal encoding.

The paper of B. Fitingof (1965) is not written clear enough, like many of the pioneering papers. A transparent setting of the problem is in L. Davisson (1973). The definitions of strong and weak universality were introduced in that paper.

Asymptotically optimal code for Bernoulli sources is developed in Krichevskii (1970).

We have discussed minimax adaptive codes in Section 3.4. The problem of adaptive encoding was set the Bayes way in Krichevskii (1968, 1970). There supposed to be a priori distribution on the set of sources. A source being observed, there appears an a posteriori distribution according to which a code is developed. Both Bayesian and minimax settings yield essentially the same results. The observation time should not be very great in both.

Move to Front code was described in Ryabko (1980, 2) under the name "stack of book". The present name was given to it in Bentley et al. (1986).

CHAPTER 4

Universal Sets of Compressing Maps

A collection of maps is a universal numerator for a group of sets, if for every set of the group there is an injective map in the collection. A collection of maps is a universal hash-set for a group of sets, if for every set of the group there is a nearly injective map in the collection. Those are generalizations of universal codes.
   We give tight asymptotic bounds on the size of universal sets.

4.1 Definition of Universal Set

Let A be a set , S be a subset of A. S will be called a source or a dictionary. Usually $A = E^n$, $n \geq 1$. The per letter entropy of S is

$$h(S) = \frac{\log |S|}{\log |A|}.$$

On the other hand, there is a set B, which is called a table. A source S is loaded into a table B. The loading factor of the table is

$$\alpha = \frac{|S|}{|B|} \qquad (4.1.1)$$

If a word x is put to the cell $f(x)$ of the table, then $f(x)$ may be called a code of x. The redundancy of such an encoding is

$$\rho = \frac{\log |B| - \log |S|}{\log |S|} \qquad (4.1.2)$$

The redundancy and the loading factor are related:

$$\alpha = |S|^{-\rho} \qquad (4.1.3)$$

Some collisions are allowed. However, their quantity should not be very great.
   We say that a set of maps $U_T(A, B, a)$ sized is a universal a-hash set for T-sized dictionaries, if for any $S \subseteq A$, $|S| = T$, there is

$$f \in U_T(A, B, a), \quad f : A \to B,$$

whose index on S does not exceed a :

$$I(f, S) \leq a.$$

Universal sets consisting only of uniform maps will be sometimes considered.

Given a set $U_T(A, B, a)$, there are two ways to solve the encoding problem. The first one is to pick a function f out of $U_T(A, B, a)$ at random. If we are lucky enough, then $I(f, S) \leq a$. If we are not, we will pick another one (rehash) and so on. The lesser $U_T(A, B, a)$ the better. We are compelled to use the random choice, if the set S is either not known beforehand or changing.

The other way is to select from $U_T(A, B, a)$ a function f, whose program length or running time are satisfactory. Such a selection may be called precomputing. A source S is transformed to a program of the function f. We are interested to have both a good function f and not very great time required to precompute S into a program of f. We use precomputing, if S is known beforehand and does not change.

An important case is $a = 0$. The set $U_T(A, B, 0)$ will be called a universal numerator and denoted, for brevity sake, by $U_T(A, B)$. For any set $S \subseteq A$ there is in $U_T(A, B)$ a map $f : A \to B$, which is injective on S : $I(f, S) = 0$.

We will deal with some other universal sets too.

A partially specified boolean function f is a map from E^n, $n \geq 1$, to $E^1 = \{0, 1\}$. The domain of f is a subset of E^n. A function f is not specified outside its domain. Say that a fully specified function (total) g and a partially specified one f agree, if for any $x \in \text{dom } f$, $f(x) = g(x)$. Fix a natural T, $0 < T < 2^n$. A set $V_T(n)$ of fully specified functions $g : E^n \to \{0, 1\}$ is said to be T-universal, if for every function f, specified on some T words of $E^n(|\text{dom } f| = T)$, there is $g \in V_T(n)$, agreeing with f. A fully specified function g is defined by a 2^n-dimensional vector, whose val x-th coordinate is $g(x)$, $x \in E^n$. So, g may be considered a vertex of the 2^n-dimensional cube E^{2^n}. A partially specified function f, $|\text{dom } f| = T$ is a $2^n - T$-dimensional subcube of E^{2^n}. Every $2^n - T$-dimensional subcube of E^{2^n} contains, or is pierced by, a vertex from $V_T(n)$. A T-universal set $V_T(n)$ may be called piercing. It is analogous to $U_T(A, B)$. Given a partial function f, $|\text{dom } f| = T$ we select (precompute) a function $g \in V_T(n)$ having a shortest program. For $x \in E^n$ we let $f(x) = g(x)$. A function g with a shortest program extrapolates f outside its domain.

Selecting a simplest function g for a partial function f we follow William Occam's Razor Principle. It states: entities should not be multiplied beyond necessity, or: it is vain to do with more what can be done with fewer. We know $f(x)$, if $x \in \text{dom } f$, and do not know $f(x)$, $x \notin \text{dom } f$. We accept Occam's Principle, take a simplest g agreeging with f and let

$$f(x) = g(x), \quad x \in \text{dom } f.$$

We also want to find a minimal $V_T(n)$.

Looking back on Chapter 3, we say, that the universal set for Bernoulli sources consists of just one map.

4.2 Covering Lemma

It is the main tool for bounding the cardinalities of universal sets. It is a variant of random coding. E. J. Nechiporuc (1965) discovered the Lemma in order to bound the cardinality of $V_T(n)$.

Lemma 4.2.1.
Let $A = \{A_1, \ldots, A_{|A|}\}$ be a set, $B = \{B_1, \ldots, B_{|B|}\}$ be a family of its subsets, every element of A belong to at least $\gamma |B|$ subset S, $0 < \gamma < 1$. Then

   1. there is a subfamily B' such that every element of A belongs to at least one subset of B',

   2. $|B'| \leq m$, where m is a number meeting the inequality

$$(1 - \gamma)^m |A| < 1.$$

In particular,

$$|B'| < \frac{\ln A}{\gamma} + 1.$$

   3. The running time to construct B' is

$$\mathrm{PT}\,(B') = \mathrm{O}\,(|B'|\,|B| \cdot \log |A| \cdot (|A| + \log |B|))$$

bitoperations.

Proof.
Make a table with $|A|$ rows and $|B|$ columns. There is 1 at the intersection of the i-th row and j-th column, if $A_i \in B_j$. There is 0 in the other case, $i = 1, \ldots, |A|$, $j = 1, \ldots, |B|$. Count the units in each column. It takes

$$|A| \cdot |B| \cdot \lceil \log |A| \rceil$$

bitoperations. Order the numbers obtained and choose a maximal one. It takes

$$|B| \cdot \lceil \log |B| \rceil \cdot \log |A|$$

bitoperations. Exclude, first, a column with the maximal quantity of units and, second, all rows having units in that column. Repeat the procedure with the new table, and so on, m times. The number m will be fixed later. The running time of the algorithm is

$$\mathrm{O}\,(m |B| \cdot \log |A| \cdot (|A| + \log |B|)).$$

There are at least $\gamma |B|$ units in each row of the table, $\gamma \cdot |B| \cdot |A|$ units in all the table, at least $\gamma |A|$ units in the column eliminated. So, there are at least $(1 - \gamma)|A|$ rows in the table after the first step.

By the same reasoning, there will be not more than

$$(1 - \gamma)^m |A|$$

rows after m steps of exclusion. The inequality

$$\ln(1 - \gamma) < -\gamma$$

shows, that after m steps there will be no rows left in the table, if

$$m > \frac{\ln|A|}{\gamma}.$$

The columns eliminated make a subfamily B' sought for. Q.E.D.

The method used is called greedy. At each step we take a subset, which covers the maximal number of elements still uncovered.

4.3 Cluster Distribution

We will use multiindex notation. If

$$a = (a_1, \ldots, a_s), \quad b = (b_1, \ldots, b_s)$$

are vectors, then

$$a! = a_1! \cdot \ldots \cdot a_s!, \quad a^b = a_1^{b_1} \cdot \ldots \cdot a_s^{b_s}, \quad |a| = a_1 + \ldots + a_s.$$

Let there be a set X, a map $f : \to [0, M - 1]$, and a dictionary S, $|S| = T$, $T > 0$. The signature of f on X is

$$\left(\left| f^{-1}(0) \right|, \ldots, \left| f^{-1}(M - 1) \right| \right).$$

A map is called uniform on X, if its signature on X is

$$u = \left(\frac{|X|}{M}, \ldots \frac{|X|}{M} \right).$$

The symbol $p_x(i)$ stands for the probability to meet a dictionary on which a map with the signature x on X has got the signature i, provided all dictionaries S, $S \subseteq X$, $|S| = |i| = T$, have the same probability. That probability depends only on the signature x of the map. The symbol $\bar{p}_x(i)$ stands for the probability to meet a map having the signatures x on X and i on a dictionary S, provided all maps f with the signature x have the same probability.

Lemma 4.3.1.

For a set X, a number T and vectors x and i, $|x| = |X|$, $|i| = T$, the probabilities $p_x(i)$ and $\bar{p}_x(i)$ are equal and

$$p_x(i) = \bar{p}_x(i) = \frac{x!\,|i|!(|\,|x| - |i|)!}{i!(x - i)!\,|x|!}.$$

Proof.

Let f be a map with a signature x. To obtain a dictionary $S \subseteq X$ on which f has a signature i one shall select i_k words from x_k words taken by f to k, $k = 0, \ldots, M-1$. Therefore, the quantity of such dictionaries equals

$$C_{x_0}^{i_0} \cdot \ldots \cdot C_{x_M}^{i_{M-1}} = \frac{x!}{i!(x-i)!}.$$

Dividing that by number $C_{|X|}^T$ of all dictionaries we obtain p_x^i.

A map f with the signature x on X and i on a dictionary S takes i_k words of S and $x_k - i_k$ words of $X§$ to k, $k = 0, \ldots, M-1$. There are $T!/i!$ ways to partition S into M subsets with cardinalities

$$(i_0, \ldots, i_{M-1}) = i.$$

Likewise, there are

$$\frac{(|X| - T)!}{(x-i)!}$$

ways to partition $X§$ into M subsets with cardinalities

$$x - i = (x_0 - i_0, \ldots, x_{M-1} - i_{M-1}).$$

Multiplying those numbers and dividing the product by the number

$$\frac{|x|!}{x!}$$

of all maps with the signature x, we obtain that $p_x(i) = \bar{p}_x(i)$. Q.E.D.

Lemma 4.3.2.

Let X be a set, T be a number, x and i be M-dimensional vectors,

$$M \geq 1, \quad \lambda = \frac{T}{|x|}x, \quad \frac{T^2}{|X|} \to 0.$$

Then the distributions $p_x(i)$ and $\bar{p}_x(i)$ are majorized by the Poisson distribution to within the factor $T!e^T T^{-T}$:

$$p_x(i) \leq \frac{\lambda^i}{i!} e^{-T} (T! e^T T^{-T}).$$

Moreover, if x is uniform, i.e., $x = (|X|/M, \ldots, |X|/M)$, and the condition

$$(1/T) \sum_{k=0}^{M-1} i_k^2 \leq C$$

holds, $C = \mathrm{const}$, then

$$p_u(i) = \frac{\lambda^i}{i!}e^{-T}(1 + \mathrm{o}\,(1))(T!e^T T^{-T}),$$

$\mathrm{o}\,(1)$ *is uniform over i,*

$$T!e^T T^{-T} \sim \sqrt{2\pi T}.$$

Proof.
From Lemma 4.3.1 we have

$$p_x(i) = \frac{T!}{i!} \cdot \frac{x_0 \dots (x_0 - i_0 + 1) \dots (x_{M-1} - i_{M-1} + 1)}{|x|\,(|X| - 1) \dots (|\,|X| - T + 1)}. \tag{4.3.1}$$

Equality (4.3.1) implies

$$p_x(i) \le \frac{T!}{i!} \left(\frac{x_0}{|X|}\right)^{i_0} \dots \left(\frac{x_{M-1}}{|X|}\right)^{i_{M-1}} \cdot \gamma_1, \tag{4.3.2}$$

where

$$\gamma_1 = \left(\left(1 - \frac{1}{|X|}\right) \dots \left(1 - \frac{T-1}{|X|}\right)\right)^{-1}.$$

If $x_0 = \dots = x_{M-1} = |X|\,/T$, then

$$p_u(i) = \frac{T!}{i!} \left(\frac{x_0}{|X|}\right)^{i_0} \dots \left(\frac{x_{M-1}}{|X|}\right)^{i_{M-1}} \gamma_2 \gamma_1, \tag{4.3.3}$$

where

$$\gamma_2 = \left(1 - \frac{T}{|X|}\right) \dots \left(1 - \frac{i_0 - 1}{|X|} \cdot T\right) \dots \left(1 - \frac{T}{|X|}\right) \dots \left(1 - \frac{i_{M-1} - 1}{|X|} \cdot T\right).$$

As it is easily seen, the inequality

$$-x \ge \ln\,(1 - x) \ge -2x$$

holds, $0 \le x \le 1/2$. Using it, we obtain

$$0 \le \ln \gamma_1 \le \frac{2}{|X|} \cdot \sum_{j=0}^{T-1} j = \frac{T(T-1)}{|X|} \tag{4.3.4}$$

The same inequality yields

$$|\ln \gamma_2| \le C \cdot \frac{T}{|X|} \sum_{k=0}^{M-1} i_k^2. \tag{4.3.5}$$

The inequality (4.3.4) and the condition $T^2/|X| \to 0$ yield

$$\gamma_1 \to 1. \tag{4.3.6}$$

The same condition (4.3.5) and the inequality

$$\sum_{k=0}^{M-1} i_k^2 \le CT$$

imply

$$\gamma_2 \to 1. \tag{4.3.7}$$

Formulae (4.3.2) and (4.3.6) imply the first Claim of the Lemma, whereas (4.3.3), (4.3.6), and (4.3.7) – the second one. The asymptotic equality

$$T! e^T \cdot T^{-T} \sim \sqrt{2\pi T}$$

is equivalent to the Stirling formula. Q.E.D.

Claim 4.3.1.

  The mathematical expectation $EI(f, S)$ of the index of a map f over all $S \subseteq X$ equals

$$\frac{T-1}{X(|X|-1)} \left(\sum_{k=0}^{M-1} x_k^2 - 1 \right),$$

where

$$x = (x_0, \dots, x_{M-1})$$

is the signature of f on X. The mathematical expectation is minimal iff f is uniform. The minimum tends to $\beta = T/M$, as $T/|X| \to 0$.

Proof.

Let

$$y = (y_0, \dots, y_{M-1}), \quad \bar{1} = (1, \dots, 1), \quad p(y) = (\bar{1} + y)^x \cdot 1/C_{|X|^T}.$$

The probability p_x^i equals the coefficient at y^i in the polynomial $p(y)$. The mathematical expectation Ei_k^2, $k = 0, \dots, M-1$ equals the coefficient at z^T in

$$y_k \frac{\partial}{\partial y_k} y_k \frac{\partial}{\partial y_k} p(y) \,|_{y_0 = y_1 = \dots = y_{M-1} = z} \ .$$

Differentiating, we obtain

$$Ei_k^2 = \frac{T \cdot x_k}{|X|} \cdot \left(\frac{(x_k - 1)(T - 1)}{|X| - 1} + 1 \right).$$

Substituting it into the definition (2.4.2) (1.2), we find $EI(f, S)$. The minimum of

$$\sum_{k=0}^{M-1} x_k^2,$$

under the condition

$$\sum_{k=0}^{M-1} x_k = |X|,$$

is assumed at the point

$$x_0 = \ldots = x_{M-1} = |X|/M.$$

Finally, we easily find

$$\lim_{T/|X|\to 0} \min I(f,S) = T/M = \beta.$$

Q.E.D.

Thus, uniform maps have got the minimal average index. They are the best ones to encode combinatorial sources. We restrict our attention to such maps only.

4.4 Large Deviations Probabilities

Given a set X, numbers T, M and a vector $x = (x_0, \ldots, x_{M-1})$,

$$p_x(I \geq b)$$

stands for the probability to meet a dictionary $S \subseteq X$, $|S| = T$, on which a function f with the signature x has got index $I(f,S) \geq b$:

$$p_x(I \geq b) = \sum p_x(i) \qquad (4.4.1)$$

$$\sum_{k=0}^{M-1} i_k^2 \geq bT + T, \quad |i| = T$$

Reciprocally, $\bar{p}_x(I \geq b)$ stands for the probability to meet a map f with the signature x, whose index $I(f,S)$ on a dictionary S is not less than b. From Lemma 4.3.1,

$$p_x(I \geq b) = \bar{p}_x(I \geq b).$$

The meaning of the symbols $p_x(I < b)$ etc. is analogous.

Lemma 4.4.1.

Let $M, T, |X|$ tend to infinity, $T/M = \beta = $ const, $T^2/|X| \to 0$. Then for any $C > 0$ there are positive constants C_1, C_2 such that

$$\text{a) } \bar{p}_u(I \leq \beta - C) = p_u(I \leq \beta - C) < e^{-C_1 T}$$

$$\text{b) } \bar{p}_u(I > \beta + C) = p_u(I > \beta + C) < e^{-C_2 \sqrt{T} \ln T}.$$

Proof.

Let

$$\lambda = (T/|X|)u = (\beta, \ldots, \beta).$$

From Lemma 4.3.2 and (4.4.1) we have

$$p_u(I \le \beta - C) = e^{-T} \sum \frac{\lambda^i}{i!}$$

$$|i| = T, \quad \sum_{k=0}^{M-1} i_k^2 \le \beta T + T - CT. \tag{4.4.2}$$

Ignoring the condition $|i| = T$ may only increase the right-hand side of (4.4.2):

$$p_u(I \le \beta - C) \le e^{-T} \sum \frac{\lambda^i}{i!} \sqrt{2\pi T}(1 + o(1))$$

$$\sum_{k=0}^{M-1} i_k^2 \le \beta T + T - CT. \tag{4.4.3}$$

Consider a sequence of independent identically distributed stochastic variables $\xi_0, \ldots \xi_{M-1}$ with the distribution

$$p(\xi_j = r^2) = e^{-\beta} \cdot \frac{\beta^r}{r!}, \quad r \ge 0, \quad 0 \le j \le M - 1. \tag{4.4.4}$$

Their first two moments are

$$E\xi_j = \beta + \beta^2, \quad E(\xi_j - E\xi_j)^2 = 4\beta^3 + 6\beta^2 + \beta. \tag{4.4.5}$$

From (4.4.4) and (4.4.5),

$$p\left(\sum_{j=0}^{M-1} \xi_j - E\xi_j \le -CT\right) = e^{-T} \sum \frac{\lambda^i}{i!}$$

$$\sum_{k=0}^{M-1} i_k^2 \le \beta T + T - CT. \tag{4.4.6}$$

a) Deviations to the left. Consider a function $\varphi(t)$,

$$\varphi(t) = e^{-t(\xi_j - E\xi_j)} = e^{-\beta} \sum_{r=0}^{\infty} e^{-t(r^2 - E\xi_j)} \frac{\beta^r}{r!}, \quad j = 0, \ldots, M - 1. \tag{4.4.7}$$

Series (4.4.7) converges uniformly, $0 \le t \le 1$. Obviously,

$$\varphi(0) = 1, \quad \varphi'(0) = 0, \quad \varphi''(t) = e^{-\beta} \sum_{r=0}^{\infty} e^{-t(r^2 - E\xi_j)} (r^2 - E\xi_j)^2 \frac{\beta^r}{r!} \le$$

$$\le e^{tE\xi_j} \cdot E(\xi_j - E\xi_j)^2 \le C', \quad 0 \le t \le 1. \tag{4.4.8}$$

Use (4.4.8) in the Taylor expansion for $\varphi(t)$:

$$\varphi(t) \leq 1 + \frac{t^2}{2} \cdot C'. \tag{4.4.9}$$

Inequality

$$Ee^{-t(\xi_j - E\xi_j)} \leq e^{\frac{C' \cdot t^2}{2}}, \quad 0 \leq t \leq 1$$

follows (4.4.7), (4.4.9) and the obvious inequality

$$e^{\frac{C' \cdot t^2}{2}} \geq 1 + C' \cdot t^2/2.$$

Now we can use the inequalities from Bernstein (1946) and from Petrov (1972, p.70, Theorem 15), yielding

$$p\left(\sum_{j=0}^{M-1}(\xi_j - E\xi_j) < -CT\right) \leq e^{-C^2 \cdot T^2/2MC'} = e^{-C'' \cdot T} \quad \text{if } CT < MC',$$

$$\leq E^{-CT/2} \quad \text{if } CT \geq MC'.$$

Letting $C_1 = \min(C'', C/2)$ we obtain from here, (4.4.6) and (4.4.3), Claim (a).

b) Deviations to the right. Here we use Theorem 2.3 from Nagaev (1979, p.765). Let

$$g(x) = C\sqrt{x}\ln x.$$

The generalized g-moment b_g of the variable ξ_j equals

$$b_g = e^{-\beta}\sum_{k=0}^{\infty} e^{C\sqrt{k^2 - E\xi_j}\ln(k^2 - E\xi_j)} \cdot \frac{\beta^k}{k!} \leq$$

$$\leq e^{-\beta}\sum_{k=0}^{\infty} e^{(C_1 k \ln k - k \ln k)(1 + o(1))}.$$

This moment exists for small enough C, $j = 0, \ldots, M - 1$.

The Taylor expansion for the exponent in the Lagrange form and the Stirling formula yield

$$p_u((\xi_j - E\xi_j) > CT) = e^{-\beta}\sum_{k^2 > E\xi_j + CT} \frac{\beta^k}{k!} \leq e^{-\beta} \cdot e^{\beta} \cdot \frac{\beta^{\sqrt{CT + E\xi_j}}}{(\sqrt{CT + E\xi_j})!} \leq$$

$$\leq e^{-C\sqrt{T}\ln T}$$

This last inequality and the existence of the g-moment suffice for the Nagaev inequality to hold. It yields

$$p_u\left(\sum_{j=0}^{M-1}\xi_j - E\xi_j > CT\right) \leq e^{-C_1\sqrt{T}\ln T}.$$

That inequality along with the inequality

$$p_u(I > \beta + C) \leq e^{-T} \cdot \sum \frac{\lambda^i}{i!}(1 + o(1))\sqrt{2\pi T}$$

$$\sum_{k=0}^{M-1} i_k^2 \geq \beta T + T + CT,$$

and the equality

$$p_u\left(\sum_{j=0}^{M-1}(\xi_j - E\xi_j) > CT\right) = e^{-T}\sum \frac{\lambda^i}{i!}$$

$$\sum_{k=0}^{M-1} i_k^2 \geq \beta T + T + CT,$$

yield Claim b) of the Lemma. Q.E.D.

Lemma 4.4.2.

Let there be $M > 0$ numbers

$$\lambda_0, \ldots, \lambda_{M-1}, \quad \lambda_0 + \ldots + \lambda_{M-1} = T, \quad T > 0, \quad \beta = T/M.$$

Then

$$\sum_{k=0}^{M-1} \lambda_i^3 \leq \left(\sum_{i=0}^{M-1} \lambda_i^2 - M\beta^2\right)^{3/2} + 3\beta\left(\sum_{i=0}^{M-1} \lambda_i^2 - M\beta^2\right) + M\beta^3.$$

Proof.

Find the maximum of the function

$$\sigma_3 = \sum_{i=0}^{M-1} \lambda_i^3$$

under the conditions

$$\sigma_1 = \lambda_0 + \ldots + \lambda_{M-1} = T, \quad \sigma_2 = \sum_{i=0}^{M-1} \lambda_i^2 = \text{const}$$

The Lagrange function is

$$\sigma_3 - \gamma_1\sigma_1 - \gamma_2\sigma_2,$$

its partial derivatives are

$$3\lambda_i^2 - 2\gamma_1\lambda_i - \gamma_2 = 0, \quad i = 0, \ldots, M - 1.$$

We have a second degree equation. Hence, variables $\lambda_0, \ldots \lambda_{M-1}$ can take no more than two values at extremal points. Denote those values by $x + \beta$ and $y + \beta$, and let the first of them be taken s times, the second one – $M - s$ times, $0 \le s \le M$. The conditions take shape

$$sx + (M - s)y = 0 \tag{4.4.10}$$

and

$$sx^2 + (M - s)y^2 + M\beta^2 = \sigma_2 = \text{const .} \tag{4.4.11}$$

For the function σ_3 we have

$$\sigma_3 = sx^3 + (M - s)y^3 + 3\beta(sx^2 + (M - s)y^2) + M\beta^3. \tag{4.4.12}$$

If either $s = 0$ or $s = M$, then the inequality of Lemma 4.4.2 turns to an equality. If $0 < s < M$, then from (4.4.10) and (4.4.11)

$$y = -\frac{s}{M - s}x, \quad x = \pm\sqrt{(M - s)/Ms}\sqrt{\sigma_2 - M\beta^2}. \tag{4.4.13}$$

Substitute (4.4.13) into (4.4.12):

$$\sigma_3 = \pm(\sqrt{\sigma_2 - M\beta^2})^3 \cdot \frac{M - 2s}{\sqrt{Ms(M - s)}} + 3\beta(\sigma_2 - M\beta^2) + M\beta^3. \tag{4:4.14}$$

The function

$$(M - 2s)/\sqrt{s(M - s)}$$

is symmetric with respect to $s = M/2$ and is decreasing, $1 \le s \le M - 1$. Hence,

$$\frac{M - 2s}{\sqrt{Ms(M - s)}} \le \frac{M - 2}{\sqrt{M(M - 1)}}. \tag{4.4.15}$$

As it is easily seen, if $M \ge 2$, then

$$\frac{M - 2}{\sqrt{M(M - 1)}} \le 1. \tag{4.4.16}$$

Equality (4.4.14) holds at the extremal points. Using (4.4.15) and (4.4.16), we obtain the inequality of the Lemma for $M \ge 2$. If $M = 1$ that inequality holds obviously. Q.E.D.

Lemma 4.4.3.

Let X be a set, M, T numbers, x an M-dimensional vector; $T^2/|X| \to 0$, $M, T, |X|$ tend to infinity, $\alpha = \text{const .}$,

$$CT^{1/2} \ge T/M = \beta \ge \alpha + CT^{-1/3}.$$

Then

$$p_x(I \le \alpha) < CT^{-1/6},$$

C does not depend on x.
Proof.
From (4.4.1) and Lemma 4.3.2

$$p_x(I \leq \alpha) \leq e^{-T} \cdot \sum \frac{\lambda^i}{i!} \cdot [T!T^{-T}e^T]$$

$$|i| = T, \quad \sum_{k=0}^{M-1} i_k^2 \leq \alpha T + T. \tag{4.4.17}$$

Here
$$\lambda = (T/|X|)x = (\lambda_0, \ldots, \lambda_{M-1}), \quad |\lambda| = T.$$

Consider a stochastic vector $\eta_0, \ldots, \eta_{M-1})$ with the probability distribution

$$p(\eta_0 = i_0^2, \ldots, \eta_{M-1} = i_{M-1}^2) = \begin{cases} \frac{\lambda^i}{i!}e^{-T}[T!T^{-T}e^T] & \text{if } |i| = T, \\ 0 & \text{otherwise.} \end{cases} \tag{4.4.18}$$

Let
$$\varphi = \varphi(y_0, \ldots, y_{M-1}) = \sum_{|i|=T} \lambda_0^{i_0} \ldots \lambda_{M-1}^{i_{M-1}} y_0^{i_0} \ldots y_{M-1}^{i_{M-1}} (T!T^{-T}) \cdot \frac{1}{i!} =$$

$$e^{\lambda_0 \cdot y_0 + \ldots + \lambda_{M-1} \cdot y_{M-1}} \cdot T!T^{-T}.$$

To obtain the mathematical expectation $E\eta_k$, we take $y_k(\frac{\partial}{\partial y_k})y_k\frac{\partial}{\partial y_k}\varphi$, then put $y_0 = \ldots = y_{M-1} = y$ and pick the coefficient at y^T out:

$$E\eta_k = T!T^{-T}\left(\lambda_k^2 \cdot \frac{T^{T-2}}{(T-2)!} + \frac{T^{T-1}}{(T-1)!}\lambda_k\right) =$$

$$= \lambda_k^2 + \lambda_k - \frac{\lambda_k^2}{T}. \tag{4.4.19}$$

Likewise

$$E\eta_k^2 = T!T^{-T}\left(\lambda_k^4\frac{T^{T-4}}{(T-4)!} + 6\lambda_k^3\frac{T^{T-3}}{(T-3)!} + 7\lambda_k^2\frac{T^{T-2}}{(T-2)!} + \lambda_k\frac{T^{T-1}}{(T-1)!}\right) \leq$$

$$\leq \lambda_k^4 + 6\lambda_k^3 + 7\lambda_k^2 + \lambda_k \tag{4.4.20}$$

and

$$E\eta_k\eta_l \leq (\lambda_k^2 + \lambda_k)(\lambda_l^2 + \lambda_l), \quad k \neq l. \tag{4.4.21}$$

Next find the dispersion D of $\sum_{k=0}^{M-1} \eta_k$:

$$D = E \left(\sum_{k=0}^{M-1} (\eta_k - E\eta_k) \right)^2 =$$

$$= \sum_{k=0}^{M-1} (E\eta_k^2 - (E\eta_k)^2) + \sum_{k \neq l} (E\eta_k \eta_l - E\eta_k E\eta_l). \qquad (4.4.22)$$

Using (4.4.19)-(4.4.21), we transform (4.4.22) into the inequality

$$D \leq \sum_{k=0}^{M-1} (4\lambda_k^3 + 6\lambda_k^2 + \lambda_k) + \frac{2}{T} \sum_{k=0}^{M-1} (\lambda_k^4 + \lambda_k^3) +$$

$$+ \frac{1}{T} \sum_{k \neq l} (\lambda_k^2 \lambda_l^2 + \lambda_k^2 \lambda_l) \leq$$

$$\leq \sum_{k=0}^{M-1} (4\lambda_k^3 + 6\lambda_k^2 + \lambda_k) + \frac{2}{T} \left(\sum_{k=0}^{M-1} (\lambda_k^2 + \lambda_k) \right)^2. \qquad (4.4.23)$$

Chebyshev inequality, (4.4.17), (4.4.18), and (4.4.19) yield

$$p_x(I \leq \alpha) \leq \frac{D}{\left(\sum\limits_{k=0}^{M-1} \lambda_k^2 - \alpha T - 1/T \sum\limits_{k=0}^{M-1} \lambda_k^2 \right)^2}. \qquad (4.4.24)$$

Substitute the inequalities of the Lemma and (4.4.23) into (4.4.24):

$$p_x(I \leq \alpha) \leq$$

$$\leq C \frac{(\sigma_2 - M\beta^2)^{3/2} + (\sigma_2 - M\beta^2)(\beta + 1) + T\beta + T + T\beta^2 + (1/T)(\sigma_2 - M\beta^2)^2}{\left((\sigma_2 - M\beta^2) + T(\beta - \alpha) - (1/T)(\sigma_2 - M\beta^2) - \beta \right)^2}$$

$$\qquad (4.4.25)$$

where $\sigma_2 = \sum\limits_{i=0}^{M-1} \lambda_i^2$. As it is known, $T\beta \leq \sigma_2 \leq T^2$ under the condition $\sum\limits_{i=0}^{M-1} \lambda_i = T$.
Hence,

$$\frac{1}{T}(\sigma_2 - M\beta^2)^2 \leq (\sigma_2 - M\beta^2)^{3/2}. \qquad (4.4.26)$$

Rewrite the conditions of the Lemma as

$$T(\beta - \alpha) \geq CT^{2/3}, \quad \beta = O\left(T^{1/12}\right). \qquad (4.4.27)$$

Two cases may present themselves, either

$$(\sigma_2 - M\beta^2)^{3/2} \leq T\beta^2$$

or

$$(\sigma_2 - M\beta^2)^{3/2} > T\beta^2.$$

In the former (4.4.25)-(4.4.27) yield

$$p_x(I \le \alpha) < C \frac{T\beta^2}{(T^{2/3} + o\,(T^{2/3}))^2} < CT^{-1/6}.$$

In the latter,

$$p_x(I \le \alpha) < C \frac{(\sigma_2 - M\beta^2)^{3/2}}{(\sigma_2 - M\beta^2)^2(1 + o\,(1))} < C \cdot T^{-1/3} \cdot \beta^{-1/3} < CT^{-1/6}.$$

Q.E.D.

4.5 Universal Numerator

A universal numerator $U_T(A, B)$ was defined in Section 4.1. For any $S \subseteq A$, $|S| = T$, there is a map $f \in U_T(A, B)$ injective on S. We use the Covering Lemma to estimate the cardinality $|U_T(A, B)|$.

Claim 4.5.1.
  Let $U_T(A, B)$ be a universal numerator. Then

$$\frac{C_{|A|}^T}{C_{|B|}^T} \left(\frac{|B|}{|A|}\right)^T \le |U_T(A, B)| \le \frac{C_{|A|}^T}{C_{|B|}^T} \left(\frac{|B|}{|A|}\right)^T \ln C_{|A|}^T.$$

The time to construct such a numerator is

$$\mathrm{PT}\,(U_T(A, B)) = \mathrm{O}\left(|B'| \cdot |B_1| \cdot \log C_{|A|}^T \cdot (C_{|A|}^T + \log|B_1|)\right),$$

where $|B'|$ is a upper bound for $|U_T(A, B)|$,

$$|B_1| = |A|! \left(\frac{|A|}{|B|}!\right)^{-|B|}.$$

Proof.
Lower bound.
  Let f be a map from A to B, f be injective on a set $S \subseteq A$. Then S contains not more than one word from each cluster $f^{-1}(b)$, $b \in B$. Denoting by $\nu(f)$ the quantity of sets, on which f is injective, we have an inequality

$$\nu(f) \le \sum \left|f^{-1}(b_1)\right| \ldots \left|f^{-1}(b_T)\right| \tag{4.5.1}$$

The sum in (4.5.1) is taken over all T-element subsets $\{b_1, \ldots b_T\}$ of B. The known Newton-Maclaurin inequality says, that symmetric sum (4.5.1) assumes its maximum, if all its multipliers are equal to each other:

$$\nu(f) \le \left(\frac{|A|}{|B|}\right)^T C_{|B|}^T \tag{4.5.2}$$

A map f is injective on $\nu(f)$sets. The number of sets is $C_{|A|}^T$. Thus,

$$|U_T(A,B)| \geq C_{|A|}^T \cdot \frac{1}{\nu(f)} = \frac{C_{|A|}^T}{C_{|B|}^T}\left(\frac{|B|}{|A|}\right)^T.$$

It is the lower bound of the Claim.

Upper bound.

For simplicity sake, let $|B|$ be a divisor of $|A|$. A map is uniform, if

$$\left|f^{-1}(b)\right| = \frac{|A|}{|B|}, \quad b \in B.$$

There are

$$|B_1| = |A|! \cdot \left(\frac{|A|}{|B|}!\right)^{-|B|}$$

uniform maps from A to B. Take a set S, $|S| = T$. Find the quantity of uniform maps f, which are injective on S. Elements of S belong to different clusters of f. There are

$$\nu_1 = |B|(|B|-1)\ldots(|B|-T+1)$$

ways to distribute those elements among the clusters. There are

$$\nu_2 = (|A|-T)! / \left(\left(\frac{|A|}{|B|}-1\right)!\right)^T \cdot \left(\frac{|A|}{|B|}!\right)^{|B|-T}$$

ways to distribute elements of $A \setminus S$ among the remaining places. Divide $\nu_1 \cdot \nu_2$ by the number $|B_1|$ of uniform maps. We obtain that for each S there are $\gamma \cdot |B_1|$ uniform maps which are injective on S,

$$\gamma = \frac{\nu_1 \nu_2}{|B_1|} = \frac{C_{|B|}^T}{C_{|A|}^T}\left(\frac{|A|}{|B|}!\right)^T.$$

Apply Lemma 4.2.1. Dictionaries S are elements of a set, uniform maps are "subsets" of a family. The upper bound of the Claim and the bound for the construction time follows from the Lemma. Q.E.D.

We will find an asymptotical form of Claim 4.5.1 if $T \to \infty$. That form depends on the relation between T and $|B|$.

Corollary 4.5.1.

Let $U_T(A,B)$ be a *universal numerator*, $\frac{T}{|A|} \to \infty$, *the loading factor* $\alpha = \frac{T}{|B|}$ *be a constant.*

Then

$$\ln|U_T(A,B)| = T\left(1 + \left(\frac{1}{\alpha}-1\right)\ln(1-\alpha)\right)(1+o(1)) + O(\ln\ln|A|)$$

If

$$\frac{\ln \ln |A|}{T} \to 0,$$

then

$$\ln |U_T(A, B)| \sim T \left(1 + \left(\frac{1}{\alpha} - 1\right) \ln (1 - \alpha)\right).$$

In particular,

$$\ln |U_T(A, B)| \sim T, \quad if \ \alpha = 1.$$

The precomputing time is

$$O \left(|B|^{|A|(1+O(1))}\right)$$

Corollary 4.5.2.
Let $U_T(A, B)$ be a universal numerator, the redundancy

$$\rho = \frac{\log |B| - \log T}{\log T}$$

be a constant.
Then

$$\ln |U_T(A, B)| \sim \frac{1}{2} T^{1-\rho}$$

The precomputing time is $O \left(|B|^{|A|(1+O(1))}\right)$
Proof.
of Corollaries 4.5.1 and 4.5.2.

The bounds of Claim 4.5.1 are given their asymptotic forms by the Stirling formula. We omit routine calculations.

Almost any dictionary S, $|S| = T$, has program complexity equal to $\ln |U_T(A, B)|$.

The upper bound of Claim 4.5.1 differs from the lower one by the multiplier $\ln C_{|A|}^T$. The omission of this multiplier does not influence asymptotical behavior of $\ln |U_T(A, B)|$, if T is great enough $\left(\frac{\ln \ln |A|}{T} \to 0\right)$.

However, if this condition is not met, then the upper bound equals the lower one multiplied by $\ln C_{|A|}^T$, which is less than $T \ln |A|$.

An example. If

$$A = E^2 = \{00, 01, 10, 11\}, \quad B = \{0, 1\} = E^1,$$

then

$$U_2(A, B) = \{f_0, f_1\}, \quad f_0 : \ 00 \to 1, \ 01 \to 1, \ 10 \to 0, \ 11 \to 0;$$
$$f_1 : \ 00 \to 1, \ 01 \to 0, \ 10 \to 1, \ 11 \to 0.$$

If

$$B = \{0, 1, 2\},$$

then

$$U_3(A, B) = \{F_0, F_1\}, \quad F_0 \text{ maps } 0 = 00 \text{ to } 0, \; 1 = 01 \to 1, \; 2 = 10 \to 0, \; 3 = 11 \to 2.$$

$$F_1 \text{ maps } 00 \to 1, \; 1 = 01 \to 0, \; 10 \to 2, \; 11 \to 0.$$

For any subset

$$S \subseteq E^2, \quad |S| = 3,$$

either F_0 or F_1 is injective on it. For instance, F_1 is injective on $\{0, 1, 2\}$, F_1 is injective on $\{0, 1, 3\}$.

Threshold function $Th_p(x_1, \ldots, x_N)$ (Sect.2.6) appears for the second time in this book. It happens to be a relative of the universal numerator.

Let

$$A = \{1, \ldots, N\},$$

N be the quantity of the variables of $Th_p(x_1, \ldots, x_N)$, $B = \{1, \ldots p\}$, $p \leq \frac{N}{2}$, p be the threshold of $Th_p(x_1, \ldots, x_N)$, $T = p$, $U_p(A, B)$ be a universal numerator, $U_p(A, B) = \{f\}$, $f : A \to B$.

Let a boolean variable x_a correspond to an element $a \in A$. A function $f \in U_p(A, B)$ divides the set A into clusters $f^{-1}(b)$, $b \in B$. A disjunction

$$\bigvee_{a \in f^{-1}(b)} x_a$$

corresponds to each cluster. A conjunction of disjunctions

$$\bigwedge_{b \in B} \bigvee_{a \in f^{-1}(b)} x_a$$

corresponds to each function f. A formula

$$\bigvee_{f \in U_p(A, B)} \bigwedge_{b \in B} \bigvee_{a \in f^{-1}(b)} x_a$$

corresponds to the set $U_p(A, B)$. For any p variables $x_{a_1}, \ldots, x_{a_p}$ there is $f \in U_p(A, B)$, which is injective on $\{a_1, \ldots, a_p\}$. Thus, those variables are placed into different disjunctions

$$\bigvee_{a \in f^{-1}(b)} x_a,$$

and the product $x_{a_1} \cdot x_{a_2} \cdot \ldots x_{a_p}$ will appear in

$$\bigwedge_{b \in B} \bigvee_{a \in f^{-1}(b)} x_a.$$

Hence,

$$Th_p(x_1, \ldots, x_N) = \bigvee_{f \in U_p(A, B)} \bigwedge_{b \in B} \bigvee_{a \in f^{-1}(b)} x_a$$

There are $|A| \cdot |U_p(A,B)|$ occurrences of variables in that formula. From the Corollary 4.5.1, where $|B| = p$, $\alpha = 1$, $|A| = N$ we get

$$|U_p(A,B)| \leq Ne^{T(1+o(1))} \qquad (4.5.3)$$

under the condition

$$\frac{\ln \ln N}{T} \to 0 <$$

and

$$|U_p(A,B)| \leq Te^{T(1+o(1))} \ln N$$

without that condition. We get

Corollary 4.5.2.

Let $\lambda_p(N)$ be the minimal complexity of formulas realizing $Th_p(x_1, \ldots, x_N)$. Then

$$CN \log N \leq \lambda_p(N) \leq Te^{T(1+o(1))} N \ln N$$

Proof.
The lower bound is proven in Claim 2.6.2. The upper bound is proven by the above correspondence

$$Th_p(x_1, \ldots, x_N) \to U_p(A,B).$$

Q.E.D.

As an example of that correspondence we take $U_3(A,B) = \{F_0, F_1\}$. F_0 subdivides $A = \{0,1,2,3\}$ into three clusters: $\{0,2\}, \{1\}, \{3\}$. A conjunction of disjunctions $(x_0 \vee x_2)x_1x_3$ corresponds to F_0. A conjunction of disjunctions $(x_1 \vee x_3)x_0x_2$ corresponds to F_1. We obtain:

$$Th_3(x_0, x_1, x_2, x_3) = (x_0 \vee x_2)x_1x_3 \vee (x_1 \vee x_3)x_0x_2.$$

4.6 Universal Hash – Sets

An universal hash-set $U_T(A,B,a)$ was defined in Section 4.1. For any dictionary $S \subseteq E^n$, $|S| = T$ it contains a uniform map, which encodes S with index a. We are going to estimate its cardinality $|U_T(A,B,a)|$. Preliminary work was done in Sections 4.3 and 4.4. An upper bound is provided by Covering Lemma 4.2.1. We estimate how many maps are there with a not very large index on a given dictionary. The family B' of the Lemma is just $U_T(A,B)$.

An easy way to obtain a lower bound is by volume argument. We estimate how many dictionaries are there on which a given map has got a not very large index. Divide the number of dictionaries by the estimate – it is just a lower bound. Sometimes that argument works. However, sometimes it gives a trivial bound. Then we use a more tricky method.

The loading factor of maps from $U_T(A, B, a)$ is $\alpha = \frac{T}{|B|}$. The following Theorem says that the cardinality $|U_T(A, B, a)|$ is either exponential or polinomial in the cardinality T of dictionaries depending on whether or not the index a is less than α.

Theorem 4.6.1.

Let A and B be sets, T and a be numbers,

$$\frac{T^2}{|A|} \to 0, \quad \lim \frac{\log T}{\log |A|} > 0,$$

the loading factor

$$\alpha = \frac{T}{|B|} = \text{const},$$

$U_T(A, B, a)$ be a universal a-hash set consisting of uniform maps. Then
a) As long as the index a is less than the loading factor α, the size of $U_T(A, B, a)$
is exponential in T :

$$C_2 \cdot T < \log |U_T(A, B, a)| < C_1 \cdot T, \quad a < \alpha$$

b) if $a = \alpha$,
$$C_3 \log T < \log |U_T(A, B, \alpha)| < \log T (1 + \text{o}(1))$$

c) if $a > \alpha$,

$$\log (\log |A| - 2 \log T) (1 + \text{o}(1)) < \log |U_T(A, B, \alpha)| <$$

$$< \frac{1}{2} \log T (1 + \text{o}(1))$$

Proof.
1. Lower bounds.
a) $a < \alpha$, $a = \alpha - C$, $C > 0$.
Use Lemma 4.4.1 with

$$\beta = \alpha, \quad X = A, \quad M = |B|.$$

We get
$$p_u(I \leq \alpha - C) \leq e^{-C_1 T}.$$

It means that the index of each uniform map is less than $\alpha - C$ for $e^{-C_1 T}$-th fraction of the set of T-sized dictionaries at the most. Therefore, the cardinality $|U_T(A, B, a)|$ can not be less than $\left(e^{-C_1 T}\right)^{-1}$. The lower bound a) is proved.
b) We prove by reductio ad absurdum that

$$|U_T(A, B, \alpha)| \geq |B|^{1/7},$$

which is equivalent to the lower bound b). Let F be a universal set,

$$F = U_T(A, B, \alpha), \quad \alpha = \frac{T}{|B|},$$

and suppose that

$$|F| < |B|^{1/7} \qquad (4.6.1)$$

Number the members of F from 0 till $|F| - 1$,

$$F = \{f_0, \ldots, f_{|F|-1}\}.$$

We will develop inductively a sequence of sets

$$A = X \supseteq X_0 \supseteq \ldots \supseteq X_{|F|}.$$

Suppose that sets

$$X_0, \ldots, X_{i-1}, \quad 0 \leq i < |F|$$

have already been determined. The set X_{i-1} is divided into, at most, $|B|$ nonempty clusters $\{f^{-1}(b)\}$, $b \in B$, with respect to the map f_i. Take the largest cluster of X_i, then the second largest one, etc,

$$\lceil |B| - |B|^{2/3} \rceil$$

clusters in all. Define X_i as the union of the clusters taken. If there are not so many clusters, define

$$X_i = X_{i-1}.$$

The quantity of those clusters is at least $\frac{1}{|B|}\lceil |B| - |B|^{2/3}\rceil$-th part of the quantity of clusters in X_{i-1}. Therefore, thanks to the way the clusters were chosen,

$$\frac{|X_i|}{|X_{i-1}|} \geq \frac{1}{|B|}(|B| - |B|^{2/3}) = 1 - |B|^{-1/3} \qquad (4.6.2)$$

Use first (4.6.1) successively and then (4.6.2). It yields

$$\frac{|X_{|F|}|}{|X_0|} \geq \frac{|X_1|}{|X_0|} \cdot \frac{|X_2|}{|X_1|} \cdots \frac{|X_{|F|}|}{|X_{|F|-1|}} \geq \left(1 - |B|^{-1/3}\right)^{|B|^{1/7}} \geq 1/2, \quad |B| \to \infty. \quad (4.6.3)$$

Inequality (4.6.3) and the condition $\lim \frac{T^2}{|A|} = 0$ of the Theorem imply that

$$\lim \frac{|X_{|F|}|}{T^2} = \infty \qquad (4.6.4)$$

Address Lemma 4.4.3 with $X_{|F|}$ playing the role of X. The range M of every $f_i \in F$ is

$$\lceil |B| - |B|^{2/3} \rceil$$

at most. For the ratio β of T to M we have, on the one hand,

$$\beta = \frac{T}{M} \geq \frac{T}{|B| - |B|^{2/3} + 1} \geq \alpha + C|B|^{1/3}, \quad \alpha = \frac{T}{|B|} \tag{4.6.5}$$

On the other hand, all maps are uniform on A. Every cluster has initially contained $\frac{|A|}{|B|}$ words. Some words were deleted when defining X_i, $i = 1, \ldots, |F|$. Hence, every f_i divides the set $X_{|F|}$ into no less than

$$\left|X_{|F|}\right| \cdot \frac{|B|}{|A|}$$

clusters. From that and (4.6.3) we have

$$\beta = \frac{T}{M} \leq \frac{T \cdot |A|}{\left|X_{|F|}\right| \cdot |B|} \leq 2\alpha \tag{4.6.6}$$

We see from (4.6.4)-(4.6.6) that all conditions of Lemma 4.4.3 are met. This Lemma yields

$$p_x(I \leq \alpha) \leq CT^{-1/6}, \tag{4.6.7}$$

where $p_x(I \leq \alpha)$ is the fraction of those dictionaries $D \subseteq X_{|F|}$, for which $I(f_i, D) \leq \alpha$; x is the signature of f_i on $X_{|F|}$, $i = 0, \ldots, |F|-1$. The probability to meet a dictionary $D \subseteq X_{|F|}$ such that there is i_0, $0 \leq i_0 < |F| - 1$, for which $I(f_{i_0}, D) < \alpha$ does not exceed

$$|F| \cdot \max_x p_x(I < a).$$

As it follows from (4.6.1) and (4.6.7), that probability goes to zero as $|B|$ goes to infinity. Therefore, there is a dictionary

$$D \subseteq X_{|F|}$$

such that

$$I(D, f_i) > \alpha$$

for all i, $0 \leq i \leq |F| - 1$ – a contraction, since F is a universal α-hash set. Consequently, (4.6.1) is not true. The lower bound b) is proven.

c) Given the index a and the loading factor α, choose a_1, $a_1 > a > \alpha$, and let

$$\gamma = \frac{\alpha}{a_1}, \quad \gamma < 1. \tag{4.6.8}$$

As it follows from the conditions

$$\lim \frac{\log T}{\log |A|} > 0 \quad \text{and} \quad \lim \frac{T^2}{|A|} = 0,$$

there is a constant $C_1 > 2$ such that

$$T^{C_1} > |A| \tag{4.6.9}$$

Let F be an a-hash-set,

$$F = \{f_0, \ldots, f_{|F|-1}\}.$$

We are going to prove that

$$|F| \geq \delta \frac{\log |A| - 2 \log T}{\log 1/\gamma}, \quad \delta = \min \left(\frac{1}{2}, \frac{1}{12(C_1 - 2)} \right), \tag{4.6.10}$$

from which Claim c) follows. Suppose, on the contrary, that

$$|F| < \delta \frac{\log |A| - 2 \log T}{\log 1/\gamma} \tag{4.6.11}$$

Develop a sequence of sets

$$A = X_0 \supseteq X_1 \supseteq \ldots \supseteq X_{|F|},$$

like it was done in b). The set X_i consists of $\lceil \gamma |B| \rceil$ largest cluster of X_{i-1}. If there are not so many clusters with respect to f_i, we define $X_i = X_{i-1}$. Obviously,

$$\left| X_{|F|} \right| \geq |A| \cdot \gamma^{|F|} \tag{4.6.12}$$

The relations (4.6.11), (4.6.12) and $\lim \frac{T^2}{|A|} = 0$ yield

$$\lim \frac{\left| X_{|F|} \right|}{T^2} = \infty \tag{4.6.13}$$

For the ratio β of T to the range of a map f_i, $0 \leq i \leq |F| - 1$, we obtain, on the one hand, from (4.6.8),

$$\beta \geq \frac{T}{\gamma |B|} \geq \frac{\alpha}{\gamma} = a_1 \geq \alpha + (a_1 - \alpha) \tag{4.6.14}$$

On the other hand, $X_{|F|}$ is divided by any $f_i \in F$ into no less than $X_{|F|} \cdot \frac{|B|}{|A|}$ clusters. Hence, from (4.6.9), (4.6.11), (4.6.12) we obtain

$$\beta \leq \frac{T \cdot |A|}{|B| \cdot \left| X_{|F|} \right|} \leq \alpha T^{1/2} \tag{4.6.15}$$

We see from (4.6.13)-(4.6.15) that all the conditions of Lemma 4.4.3 are met. Therefore,

$$p_x(I \leq \alpha) \leq CT^{-1/6}.$$

It yields, along with (4.6.11) and (4.6.9), that there is a dictionary $D \subsetneq X_{|F|}$, on which the index of each $f_i \in F$ is greater than α – a contradiction. Consequently, (4.6.11) does not hold. The lower bound c) is proven.

2. <u>Upper bound.</u>

a) Index a is less than the loading factor α. The cardinality of an α-hash set $U_T(A, B, \alpha)$ is less than the cardinality of the corresponding 0-hash set (numerator):

$$|U_T(A, B, \alpha)| \le |U_T(A, B, 0)| = |U_T(A, B)| .$$

An exponential upper bound for $|U_T(A, B)|$ is given in Corollary 4.5.1. That bound suits us.

b) The loading factor equals the index, $a = \alpha$. The index is related to χ^2-statistic by (2.4.4):

$$I(f, S) = \frac{\chi^2 + T}{|B|} - 1.$$

Medvedev (1970) showed that χ^2-test is normally distributed under the conditions of Theorem 4.6.1. Hence, the uniform maps f, whose index on a dictionary S, $|S| = T$, does not exceed a, constitute at least γ-th fraction of all such maps, where

$$\gamma \ge \frac{1}{2} - \varepsilon, \qquad (4.6.16)$$

ε is arbitrarily small, $|A| \to \infty$. We are in the domain of the Covering Lemma. Dictionaries S, $|S| = T$, $S \subseteq A$ are elements of a set, uniform maps play the role of covering subsets. The quantity of dictionaries is $C_{|A|}^T$. There is a universal set $U_T(A, B, \alpha)$, whose cardinality $|U_T(A, B, \alpha)|$ equals m, where

$$(1 - \gamma)^m C_{|A|}^T < 1 \qquad (4.6.17)$$

Apply the Stirling formula to (4.6.17), using the conditions of the Theorem and (4.6.16). We obtain, that there is a universal set $U_T(A, B, \alpha)$, $\frac{T}{|B|} = \alpha$, such that

$$|U_T(A, B, \alpha)| \le CT(\log |A| - \log T)$$

It is equivalent to the upper bound b), because

$$\lim \frac{\log T}{\log |A|} > 0.$$

c) The index a is greater than the loading factor α.

The uniform maps whose index on a dictionary S, $|S| = T$, does not exceed a, constitute at least γ-th fraction of all such maps, where

$$\gamma \ge 1 - e^{-C\sqrt{T} \ln T}$$

by the second case of Lemma 4.4.1. The Covering Lemma, the Stirling formula and the conditions of the Theorem yield

$$|U_T(A, B, \alpha)| \le \frac{1}{2} \log T (1 + o(1)).$$

The proof of the Theorem is completed.

4.7 Piercing Sets and Independent Sets

A set $V_T(n)$ of vertices of the 2^n-dimensional cube E^{2^n} was called piercing, if every $2^n - T$ dimensional subcube contains at least one vertex of $V_T(n)$. There are $2^{2^n - T}$ vertices in every $2^n - T$-dimensional subcube. There are

$$C_{2^n}^T \cdot 2^T = O\left(2^{2^n}\right)$$

such subcubes.

Every vertex of E^{2^n} belongs to $C_{2^n}^T$ subcubes, which is $\gamma = 2^{-T}$-th fraction. So, by the Covering Lemma,

$$2^T \cdot \ln C_{2^n}^T 2^T$$

vertices are enough to pierce every subcube. On the other hand, less than 2^T vertices are not enough. We come to

Claim 4.7.1.

The cardinality of a piercing set $V_T(n)$ meets the inequalities

$$2^T \le |V_T(n)| \le 2^T \cdot n \cdot (T + 1) \cdot \ln 2$$

The time to precompute such set is $O\left(2^{2^n}\right)$.

Table 4.7.1 shows a set $V_3(2)$. It consists of eight vertices of 4-dimensional cube. They are numbered from 0 to 7. Their coordinates are numbered from 0 to 4. Every vertex may be considered as a boolean function of two variables.

Table 4.7.1.

A set $V_3(2)$. For any partial boolean function f of two variables, $|\text{dom } f| = 3$, there is a column agreeing with f.

vectors	The number of a function (vertex) in $V_3(2)$							
from E^2	000	001	010	011	100	101	110	111
00	0	1	1	1	0	0	0	1
01	0	1	0	0	1	1	0	1
10	0	0	1	0	0	1	1	1
11	0	0	0	1	1	0	1	1

A piercing set was interpreted another way in Becker, Simon (1988). Make for such a set a $2^n x |V_T(n)|$-table, whose rows correspond to vertices of E^n, columns - to the functions from $V_T(n)$. There is $f(x)$ at the intersection of a row x and a column f, $x \in E^n$, $f \in V_T(n)$. A row may be considered as a characteristic vector of a subset A of $V_T(n)$. If an element of a row is 1, then the corresponding function is included in A. Let $A_1, A_2, \ldots$ be subsets of $V_T(n)$, which correspond to rows of the table. Take any T of them, $A_{i_1}, \ldots A_{i_T}$. Let B_{i_k} be either A_{i_k} or its complement, $k = 1, \ldots, T$. As it follows from the definition of $V_T(n)$, the intersection

$$B_{i_1} \cap \ldots \cap B_{i_k}$$

is nonempty. It is equivalent to the condition that every $2^n - T$-dimensional subcube contains at least one vertex of $V_T(n)$.

A family of subsets of a set is called T-independent, if for any T subsets $A_{i_1}, \ldots, A_{i_T}$ the intersection $B_{i_1} \cap \ldots \cap B_{i_k}$ is nonempty, B_{i_k} being either A_{i_k} or its complement. Thus, a piercing set produces a T-independent family.

Rewrite Claim 4.7.1 as inequality

$$n \geq |V_T(n)| \cdot 2^{-T} \cdot T^{-1} \cdot C \geq 2^{-CT} \cdot |V_T(n)|, \quad C = \text{const}, \qquad (4.7.1)$$

Let

$$N = 2^n, \quad |V_T(n)| = M.$$

Rewrite (4.7.1):

$$\log N \geq 2^{-Ct} \cdot M.$$

It means that there is a $2^{C_T M}$-sized family of T-independent subset of a M-sized set, where $C_T = 2^{-CT}$.

NOTES

E. Nechiporuk (1965) introduced piercing sets and proved the Covering Lemma. The lower bound of Lemma is obtained by volume argument: how many subsets are required to cover a set. The upper bound is obtained by random coding argument. The volume argument does not work for universal hash-sets, so the proof is more tricky.

A relation between universal numerators and threshold functions was discovered by L. S. Hasin (1969). It enabled him to get short threshold formulas. T-independent families are constructed by Kleitman and Spencer (1973).

CHAPTER 5

Elementary Universal Sets

Lexicographic, digital, enumerative, Galois linear and polynomial methods of retrieval are called elementary. The first three of them are widely known. The Galois methods are defined in Sections 5.2 and 5.3. We present those retrieval algorithms in a non standard way: each of them determines a family of maps, which constitute a universal hash-set, a universal numerator in particular.

A universal hash-set is characterized by its size, precomputing time and the running time of its maps. To precompute means to select an injective on a dictionary map from a universal set. The logarithm of the size of the set is the program complexity.

The universal sets of Chapter 4 were constructed by exhaustive search. Such a tiresome procedure yields sets of optimal size, as a rule. The sizes of elementary sets are far from the minimum. On the other hand, they are described explicitly, which means that the precomputing time is reasonably small.

Some elementary retrieval algorithms are improved as compared with their conventional variants. So, the running time of lexicographic search of Section 5.7 is $\log T \cdot \log n$, instead of usual $n \cdot \log T$, n being the wordlength, T being the dictionary size. Two-step digital search has a much shorter program than the usual one-step digital search.

Polynomial hashing is quite easy to implement.

Consider unordered search as a point of departure. A dictionary $S \subseteq A$ is itself a program of retrieval. The set A is assumed to coincide with E^n, $n = \log |A|$, $|S| = T$. Hence, the unordered retrieval defines a universal numerator

$$U_T(A, B), \quad |B| = T, \quad |U_T(A, B)| = |A|^T.$$

The precomputing time is zero.

Next we go to not so obvious methods of retrieval. Their precomputing time is greater, their size is less. Some of those methods are based on Theorem 5.1.1.

String matching is an application of the polynomial hashing.

5.1 Universal Sets Based on Codes

Let A and B be sets, $\Phi = \{\varphi_1, \ldots, \varphi_{|\Phi|}\}$ be a set of maps from A to B. The tensor product $\otimes \Phi$ of maps from Φ takes a word $x \in A$ to the concatenation of words

$\varphi_1(x)\varphi_2(x)\ldots\varphi_{|\Phi|}(x)$. The Hamming distance $\rho(w_1, w_2)$ between two words in the alphabet B equals the number of mismatches between w_1, w_2. In particular,

$$\rho(\otimes\Phi x, \otimes\Phi y) = \sum_{i=1}^{|\Phi|} \rho(\varphi_i(x), \varphi_i(y)) \tag{5.1.1}$$

Here

$$\rho(\varphi_i(x), \varphi_i(y)) = 1, \quad \text{if } \varphi_i(x) = \varphi_i(y),$$

in the other case

$$\rho(\varphi_i(x), \varphi_i(y)) = 0.$$

A set is a code with distance r, if the distance between every two its words is not less than r.

Theorem 5.1.1.

Let $\Phi = \{\varphi_1, \ldots \varphi_{|\Phi|}\}$ be a set of maps from A to B, $S \subseteq A$, $\otimes\Phi$ be a r-distance code on S, $I_1 = I(\varphi_1, S)$ be the index of φ_1 on S.

Then there is $\varphi_{i_0} \in \Phi$ such that the index $I(\varphi_1\varphi_{i_0}, S)$ of the map $\varphi_i\varphi_{i_0}$, that takes $x \in S$ to the concatenation $\varphi_1(x)\varphi_{i_0}(x)$, is not greater than $I_1 \cdot \left(1 - \frac{r}{|\Phi|}\right)$:

$$I(\varphi_1\varphi_{i_0}, S) \leq I_1 \cdot \left(1 - \frac{r}{|\Phi|}\right)$$

Proof.

Let $n_j(\alpha)$ be the cardinality of the coimage of $\alpha \in B$ under $\varphi_j \in \Phi$:

$$n_j(\alpha) = \left|\varphi_j^{-1}(\alpha)\right|.$$

Denote by $n_{1i}(\alpha, \beta)$ the number of words of S, which are taken by φ_1 to α and by φ_i to β :

$$n_{1i}(\alpha, \beta) = \left|\varphi_1^{-1}(\alpha) \cap \varphi_i^{-1}(\beta)\right|.$$

As it is clear from the definitions, for every i,

$$n_1(\alpha) = \sum_{\beta \in B}^{2 \leq i \leq |\Phi|} n_{1i}(\alpha, \beta) \tag{5.1.2}$$

Square (5.1.2):

$$n_1^2(\alpha) = \sum_{\beta} n_{1i}^2(\alpha, \beta) + 2 \sum_{\beta_1, \beta_2, \beta_1 \neq \beta_2} n_{1i}(\alpha, \beta_1)n_{1i}(\alpha, \beta_2) \tag{5.1.3}$$

In (5.1.3) $\alpha, \beta, \beta_1, \beta_2$ belong to B, $2 \leq i \leq |\Phi|$.

Consider the sum

$$\sum_{x, y; \varphi_1(x)=\varphi_1(y)=\alpha} \rho(\varphi_i(x), \varphi_i(y))$$

There are $n_1^2(\alpha)$ couples x, y, for which $\varphi_1(x) = \varphi_1(y) = \alpha$. Exactly

$$\sum_{\beta \in B} n_{1i}^2(\alpha, \beta)$$

of them do not contribute to the sum, $2 \le i \le |\Phi|$. Equalities (5.1.2) and (5.1.3) yield

$$2 \sum_{\beta_1, \beta_2, \beta_1 \ne \beta_2} n_{1i}(\alpha, \beta_1) n_{1i}(\alpha, \beta_2) = \sum_{x,y,\varphi_1(x)=\varphi_1(y)=\alpha} \rho(\varphi_i(x), \varphi_i(y)). \qquad (5.1.4)$$

By the definition,

$$I(\varphi_1, S) = I_1 = \frac{1}{|S|} \sum_{\alpha} n_1^2(\alpha) - 1 \qquad (5.1.5)$$

Similarly, for every i, $2 \le i \le |\Phi|$,

$$I_2 = \frac{1}{|S|} \sum_{\alpha, \beta} n_{1i}^2(\alpha, \beta) - 1 \qquad (5.1.6)$$

Consider a triple sum and change in it the summation order:

$$\sum_{i=2}^{|\Phi|} \sum_{\alpha \in B} \sum_{x,y;\varphi_1(x)=\varphi_1(y)=\alpha} \rho(\varphi_i(x), \varphi_i(y)) =$$

$$= \sum_{a \in B} \sum_{i=2}^{|\Phi|} \sum_{x,y;\varphi_1(x)=\varphi_1(y)=\alpha} \rho(\varphi_i(x), \varphi_i(y)) = \sum_{\alpha \in B} \sum_{i=1}^{|\Phi|} \sum_{x,y;\varphi_1(x)=\varphi_1(y)=\alpha} \rho(\varphi_i(x), \varphi_i(y))$$

$$(5.1.7)$$

Rewrite (5.1.7) using formula (5.1.1):

$$\sum_{i=2}^{|\Phi|} \sum_{\alpha \in B} \sum_{x,y;\varphi_1(x)=\varphi_1(y)=\alpha} \rho(\varphi_i(x), \varphi_i(y)) = \sum_{\alpha \in B} \sum_{x,y;\varphi_1(x)=\varphi_1(y)=\alpha} \rho(\otimes \Phi x, \otimes \Phi y) \qquad (5.1.8)$$

By the condition of Theorem 5.1.1, the words $\otimes \Phi x$ make a r-distance code. Thus,

$$\sum_{\alpha \in B} \sum_{x,y;\varphi_1(x)=\varphi_1(y)=\alpha} \rho(\varphi_i(x), \varphi_i(y)) \ge r \sum_{\beta \in B} n_1(\alpha)(n_1(\alpha) - 1)$$

As it follows from (5.1.8) and (5.1.9), there is a number i_0, $1 \le i_0 \le |\Phi|$, such that

$$\sum_{\alpha \in B} \sum_{x,y;\varphi_1(x)=\varphi_1(y)=\alpha} \rho(\varphi_{i_0}(x), \varphi_{i_0}(y)) \ge \frac{r}{|\Phi|} \sum n_1(\alpha)(n_1(\alpha) - 1) \qquad (5.1.10)$$

Substitute (5.1.3) and (5.1.4) into (5.1.10):

$$\sum_{\alpha \in B} n_1^2(\alpha) - \sum_{\alpha, \beta} n_{1i_0}^2(\alpha, \beta) \ge \frac{r}{|\Phi|} \sum_{\alpha \in B} n_1(\alpha)(n_1(\alpha) - 1) \qquad (5.1.11)$$

It remains to apply definitions (5.1.5) and (5.1.6) to (5.1.11) in order to obtain the claim of the Theorem. Q.E.D.

Corollary 5.1.1.
   A set of maps

$$\Phi = \{\varphi_1, \dots \varphi_{|\Phi|}\}, \quad \varphi_i : A \to B,$$

is a universal $U_T(A, B, a)$ *- set, where*

$$a \leq (T - 1)\left(1 - \frac{r}{|\Phi|}\right),$$

if $\otimes\Phi$ *constitute a* r*-distance code.*

Proof.
The maximal value of the index on a T-sized dictionary is $T - 1$. Take the empty map as φ_1. We get the Corollary.

   Take a dictionary S, which consists of n-length binary words: $S \subseteq E^n$, $n \geq 1$. Let $S(l, \sigma)$ stand for the subset of those words S, that have got σ at the l-th place, $\sigma = 0, 1$, $l = 1, \dots, n$. Obviously,

$$|S(l, 0)| + |S(l, 1)| = |S| = T. \tag{5.1.12}$$

Corollary 5.1.2.
   For every dictionary S, $|S| = T \geq 2$, *an inequality*

$$\max_{1 \leq i \leq n} \min_{\sigma = 0, 1} |S(i, \sigma)| \geq \frac{|S|}{4n}$$

holds.

Proof.
Let φ_i maps a word x into its i-th letter, $i = 1, \dots, n$, $\Phi = \{\varphi_1, \dots, \varphi_n\}$. Then

$$\otimes\Phi x = x,$$

The set $\otimes\Phi x$ is a 1 - distance code. Corollary 5.1.1 guarantees the existence of a map φ_j, $1 \leq j \leq n$ such that

$$I(\varphi_j, S) \leq (T - 1)\left(1 - \frac{1}{n}\right) \tag{5.1.13}$$

By the definition of φ_j we get

$$I(\varphi_j, S) = \frac{1}{T}\left(|S(j, 0)|^2 + |S(j, 1)|^2\right) - 1 \tag{5.1.14}$$

Square (5.1.12) and subtract (5.1.13) from the result:

$$2|S(j, 0)| \cdot |S(j, 1)| \geq \frac{T(T - 1)}{2n} \tag{5.1.15}$$

Use (5.1.12) in (5.1.15). We get

$$\frac{|S(j,\sigma)|}{T} \geq \frac{|S(j,\sigma)|}{T}\left(\frac{|S(j,\bar{\sigma})|}{T}\right) =$$

$$= \frac{|S(j,0)|\,|S(j,1)|}{T^2} \geq \frac{1}{2n}\left(1 - \frac{1}{T}\right) \geq \frac{1}{4n} \qquad (5.1.16)$$

where $\bar{\sigma} = 0$, if $\sigma = 1$, and $\bar{\sigma} = 1$, if $\sigma = 0$. We can see from (5.1.16) that there is a number j for which

$$|S(j,\sigma)| \geq \frac{T}{4n},$$

Q.E.D.

The meaning of the Corollary is that for a large enough dictionary S there is a position j, which divides S into two parts $S(j,0)$ and $S(j,1)$ of nearly equal size. We consider n as negligible in comparison with T.

5.2 Linear Galois Hashing

First, a short reminder about main facts of the Galois theory.

A set with two operations: multiplication and addition is called a field. Those operations are commutative and associative. The distributive law is valid. There are elements zero and unit. For every x, $x \neq 0$, there are $-x$ and x^{-1} such that $x + (-x) = 0$, $x \cdot x^{-1} = 1$. The set of residues modulo a prime is a field. For instance, the set of residues modulo 2 consists of two elements, 0 and 1, $0 + 0 = 1 + 1 = 0$. That field is denoted by $GF(2)$.

A polynomial is irreducible, if it is not a nontrivial product of two polynomials. For any natural μ there is a irreducible polynomial, whose degree is μ.

Take an irreducible polynomial $g(x)$ of degree μ, whose coefficients belong to $GF(2)$.

The set of residues modulo $g(x)$ constitutes a field denoted by $GF(2^\mu)$. There are 2^μ elements in $GF(2^\mu)$.

There is an element x_0 of $GF(2^\mu)$, which is called primitive. Every nonnul element x of $GF(2^\mu)$ is a power of x_0. That power is called the index of x.

Let $g(x) = x^2 + x + 1$. It is irreducible. Residues modulo $g(x)$ are $0 + 0 \cdot x$, $1 + 0 \cdot x$, $0 + 1 \cdot x$, $1 + x = x^2$. Every element of $GF(2^\mu)$ is a $\mu - 1$ – degree polynomial, whose set of coefficients is a μ-dimensional vector. Thus, $g(x)$ sets an isomorfism between $GF(2^\mu)$ and E^μ. A primitive element of $GF(2^2)$ is x. The indices are: ind $1 = 0$, ind $x = 1$, ind $(1 + x) = 2$.

Let $g(x) = x^3 + x + 1$. It is irreducible. A primitive element of $GF(2^3)$ is x. We have:

$$x^0 = (1,0,0), \quad x^1 = (0,1,0), \quad x^2 = (0,0,1), \quad x^3 = 1 + x = (1,1,0),$$

$$x^4 = x^2 + x = (0,1,1), \quad x^5 = (1,1,1), \quad x^6 = (1,0,1), \quad x^7 = (0,0,1).$$

Table 5.2.1 shows indices for $GF(2^8)$.

Table 5.2.2 is its inverse. It shows exponents for the same field.

Table 5.2.1.

Indices for the Galois field $GF(2^8)$															
-	0	1	157	2	59	158	151	3	53	69	132	159	70	152	216
4	118	54	38	61	47	133	227	160	181	71	210	153	34	217	16
5	173	119	221	55	43	39	191	62	88	48	83	134	112	228	247
161	28	182	20	72	195	211	242	154	129	35	207	218	80	17	204
6	106	174	164	120	9	222	237	56	67	44	31	40	109	192	77
63	140	89	185	49	177	84	125	135	144	113	23	229	167	248	97
162	235	29	75	183	123	21	95	73	93	196	198	212	12	243	200
155	149	130	214	36	225	208	14	219	189	81	245	18	240	205	202
7	104	107	65	175	138	165	142	121	233	10	91	223	147	238	187
57	253	68	51	45	116	32	179	41	171	110	86	193	26	78	127
64	103	141	137	90	232	186	146	50	252	178	115	85	170	126	25
136	102	145	231	114	251	24	169	230	101	168	250	249	100	98	99
163	105	236	8	30	66	76	108	184	139	124	176	22	143	96	166
74	234	94	122	197	92	199	11	213	148	13	224	244	188	201	239
156	254	150	58	131	52	215	69	37	117	226	46	209	180	15	33
220	172	190	42	82	87	246	111	19	27	241	194	206	128	20	80

An example: for $x = 129 = 1000\ 0001$, ind x is at the intersection of 2-th column and 9-th row. Ind $x = 104 = 0110\ 1000$.

Table 5.2.2.

Exponents for the Galois field $GF(2^8)$															
1	2	4	8	16	32	64	128	195	69	138	215	109	218	119	238
31	62	124	248	51	102	204	91	182	175	157	249	49	98	196	75
150	239	29	58	116	232	19	38	76	152	243	37	74	148	235	21
42	84	168	147	229	9	18	36	72	144	227	5	10	20	40	80
160	131	197	73	146	231	13	26	52	104	208	99	198	79	158	255
61	122	244	43	86	172	155	245	41	82	164	139	213	105	210	103
206	95	190	191	189	185	177	161	129	193	65	130	199	77	154	247
45	90	180	171	149	233	17	34	68	136	211	101	202	87	174	159
253	57	114	228	11	22	44	88	176	163	133	201	81	162	135	205
89	178	167	141	217	113	226	7	14	28	56	112	224	3	6	12
24	48	96	192	67	134	207	93	186	183	173	153	241	33	66	132
203	85	170	151	237	25	50	100	200	83	166	143	221	121	242	39
78	156	251	53	106	212	107	214	111	222	127	254	63	126	252	59
118	236	27	54	108	216	115	230	15	30	60	120	240	35	70	140
219	117	234	23	46	92	184	179	165	137	209	97	194	71	142	223
125	250	55	110	220	123	246	47	94	188	187	181	169	145	22	-

The Fermat theorem holds:

$$x^{2^\mu - 1} = 1, \quad x \neq 0, \quad x \in GF(2^\mu).$$

We will use Theorem 5.1.1 to develop two universal sets.

Given two elements,

$$a = a_{\mu-1}x^{k-1} + \ldots + a_0, \quad b = b_{\mu-1}x^{\mu-1} + \ldots + b_0$$

of $GF(2^\mu)$ there are two ways to find their product xy. The first one is straightforward. Multiply a by b as polynomials. It takes $O(\mu^2)$ operations. Divide the result by $g(x)$. It takes $O(\mu^2)$ more operations, $O(\mu^2)$ operations in all.

The other way is to take a table of indices for $GF(2^\mu)$ and its inverse. Find ind x, ind y, ind x + ind $y \pmod{2^{\mu-1}}$ and then xy by the inverse table. It takes $O(\mu)$ operations and requires two tables of $\mu \cdot 2^\mu$ bits size.

Take each pair of $\mu - 1$ - degree polynomials and divide one of them by the other. In such a way we can find an irreducible polynomial. The time consumption is $O(\mu^2 \cdot 2^{2\mu})$ bitoperations ($2^{2\mu}$ pairs, $O(\mu^2)$ operations to divide).

Take an element g of $GF(2^\mu)$ and multiply it by itself $2^\mu - 1$ times ($O(2^\mu \cdot \mu^2)$)bitoperations). Order the set $\{g^k\}$, $k = 1, \ldots, 2^\mu - 1$ of μ-length vectors obtained ($2^\mu \cdot \mu^2$ bitoperations). Verify whether all vectors are different. If they are, then g is primitive. If they are not, take another element of $GF(2^\mu)$. It takes $O(\mu^2 \cdot 2^{2\mu})$ bitoperations to find a primitive element of $GF(2^\mu)$.

We obtain a table of indices along with a primitive element of $GF(2^\mu)$.

Take the set E^n of all binary n-length words, $n \geq 1$. Let μ be a divisor of n. Every word w is the concatenation of $\lceil \frac{n}{\mu} \rceil$ subwords, each being of μ bits length:

$$w = w_1 w_2 \dots w_{\lceil n/\mu \rceil}, \quad |w_i| = \mu, \quad i = 1, \dots, \frac{n}{\mu}.$$

Take a n-length vector b,

$$b = b_1 \dots b_{\lceil n/\mu \rceil}$$

and define a linear map

$$\varphi_b : E^n \to E^\mu,$$

$$\varphi_b(w) = b_1 w_1 + \dots b_{\lceil n/\mu \rceil} w_{\lceil n/\mu \rceil} \tag{5.2.1}$$

Operations are done in $GF(2^\mu)$. Formula (5.2.1) describes a family $\Phi = \{\varphi_b\}$ of maps, $|\Phi| = 2^n$.

Suppose that there are two vectors w' and w'' such that $\varphi_b(w') = \varphi_b(w'')$. Then

$$\sum_{i=1}^{\lceil n/\mu \rceil} b_i(w_i' + w_i'') = 0 \tag{5.2.2}$$

Vectors w' and w'' being different, there is i_0, $1 \leq i_0 \leq \frac{n}{\mu}$ such that

$$w_{i_0}' + w_{i_0}'' \neq 0.$$

Fix vectors $b_1, \dots, b_{n/\mu}$, except b_{i_0}. We can find b_{i_0} from equation (5.2.2). Thus, there are $2^{(n/\mu-1)\mu}$ different n-dimensional vectors b meeting (5.2.2). Vectors $\otimes \Phi w_1$ and $\otimes \Phi w_2$ have not less than

$$r = 2^n - 2^{\mu(n/\mu-1)} = 2^n \cdot \left(1 - 2^{-\mu}\right)$$

different positions. We come to

Claim 5.2.1.

For every dictionary S, $|S| = T$, $S \subseteq E^n$, $n > 1$, and an integer μ there is a linear map $\varphi_b : E^n \to E^\mu$ such that

$$I(\varphi_b, S) \leq (T-1) \cdot 2^{-\mu}.$$

The time required to find that map is $\mathrm{O}\left(2^n \cdot 2^\mu \cdot n\right)$ given a table of indices or $\mathrm{O}\left(2^n \cdot 2^\mu \cdot n\mu\right)$ without such a table.

In other words, the set of linear maps is a universal set $U_T\left(E^n, E^\mu, (T-1)/2^\mu\right)$. The Claim follows Corollary 5.1.1. The map is selected by exhaustive search.

The loading factor of maps φ_b is $T/2^\mu$. That equals asymptotically the index. The cardinality of $U_T(E^n, E^\mu, (T-1)/2^\mu)$ is 2^n. We have

$$C \log T < \log \left| U_T(E^n, E^\mu, \frac{T-1}{2^\mu}) \right| = n, \quad T = Cn.$$

We can say, that the cardinality of the universal set developed by Claim 5.2.1 is close enough to the lower bound b) of Theorem 4.6.1.

Given a dictionary S, an element b, for which

$$I(\varphi_b, S) < \frac{T-1}{2^\mu},$$

is taken either at random or by exhaustive search.

5.3 Polynomial Galois Hashing

Subdivide a word $w \in E^n$ into μ-length subwords just as it was done for the linear hashing. Take an element $b \in GF(2^\mu)$ and define a polynomial map $f_b : E^n \rightarrow E^\mu$

$$f_b(w) = w_1 + w_2 b + \ldots + w_{\frac{n}{\mu}} b^{\frac{n}{\mu}-1} \tag{5.3.1}$$

Formula (5.3.1) describes a set Φ, $|\Phi| = 2^\mu$, of maps from E^n to E^μ. We denote that set by $G(n, \mu)$.

There are two functions, f_0 and f_1 in $G(2, 1)$: $f_0(w) = w_1$, $f_1(w) = w_1 + w_2$. There are four function in $G(4, 2)$. Represent a vector w by the corresponding number val w in decimal notations: $10 = 2$, $11 = 3$, $1111 = 15$, etc. Every function f is represented by a list of 16 numbers. There is $f(w)$ at the val w-th place. Table 5.3.1 shows those lists.

Table 5.3.1.

place	0	1	2	3	4	5	6	7	8	9	10	11	12	13	14	15
f_0	0	0	0	0	1	1	1	1	2	2	2	2	3	3	3	3
f_1	0	1	2	3	1	0	3	2	2	3	0	1	3	2	1	0
f_2	0	2	3	1	1	3	2	0	2	0	1	3	3	1	0	2
f_3	0	3	1	2	1	2	0	3	2	1	3	0	3	0	2	1

The table is made by formula (5.3.1). For example, if $b = 3$, $w = 10$, then in binary notation

$$3 = 11, \quad 10 = 1010,$$

$$f_3(10) = f_{11}(1010) = (1, 0) + (11)(1, 0) =$$

$$= x + (x+1)x = x^2 = x + 1 (\text{mod } x^2 + x + 1) = (1, 1) = 3.$$

A polynomial (5.3.1) has $\frac{n}{\mu} - 1$ zeroes at most. Hence, the Hamming distance between two words $\otimes \Phi w'$ and $\otimes \Phi w''$ is $2^\mu - \frac{n}{\mu} + 1$. By Theorem 5.1.1, for any $S \subseteq E^n$ there is $b \in GF(2^\mu)$ such that

$$I(f_b, S) \le \frac{T}{2^\mu} \left(\frac{n}{\mu} - 1 \right) \tag{5.3.2}$$

In other words, $G(n, \mu)$ is a $U_T(E^n, E^\mu, \frac{T}{2^\mu}\left(\frac{n}{\mu} - 1\right))$ universal set. Its cardinality is 2^μ.

If

$$\frac{T}{2^\mu}\left(\frac{n}{\mu} - 1\right) < \frac{2}{T}, \qquad (5.3.3)$$

in particular, if

$$\mu = \lceil 2\log T + \log n \rceil,$$

then, by Lemma 2.4.2, for any S, $|S| = T$, the set $G(n, \mu)$ of polynomial maps contains an injective functions. In other words, if μ meets (5.3.3), then the set of polynomial maps $G(n, \mu)$ is a universal numerator. The set $G(4, 2)$ is a universal numerator of 3-element subsets of E^4. It can be verified directly. The functions f_1, f_2, f_3 are injective on $S = \{0, 1, 2\}$, f_2 – on $\{0, 3, 5\}$, f_3 – on $\{6, 7, 8\}$, etc.

To calculate a polynomial $f_b(w)$ we use the Horner scheme:

$$f_b(w) = \left(\left(w_{\frac{n}{\mu}}b + w_{\frac{n}{\mu}-1}\right)b + \ldots\right).$$

It takes $O\left(\frac{n}{\mu} \cdot \mu^2\right) = O(n\mu)$ bitoperations, if the multiplication is done straightforwardly. It takes $O(n)$ operations and $O(\mu 2^\mu)$ bits of memory, if the multiplication is done via a table of indices.

We can choose a good polynomial f_b for a given dictionary S by exhaustive search. Take an element $b \in GF(2^\mu)$ and compute $f_b(w)$ for all $w \in S$. The computation time is Tn, provided a table of indices. Order coordinates of the vector $f_b(w)$, $w \in S$. It takes

$$O(T \cdot \log T \cdot n) = O(Tn \log T)$$

bitoperations. Given the ordered set of coordinates it is easy to find the index $I(f_b, S)$. Finally, select a $b \in GF(2^\mu)$ for which the index is minimal. We have used $O(Tn \log T \cdot 2^\mu)$ bitoperations.

We can formulate

Claim 5.3.1.

Let A and B be sets, T be natural, $|B| \geq T^2 \log |A|$. There is a universal numerator $U_T(A, B)$, $|U_T(A, B)| \leq T^2 \log |A|$. It takes $O(\log |A|)$ units of time to compute a function $f \in U_T(A, B)$. It takes $O(T^3) \log T (\log |A|)^2$ units of time to precompute for a dictionary $S \subseteq A$ a function $f \in U_T(A, B)$, which is injective on S.

Every function $f \in U_T(A, B)$ is a polynomial f_b, where

$$|b| = \lceil 2\log T + \log n \rceil.$$

The word b is a computing program of f.

There is another method to choose a good polynomial f_b. It is both less obvious and less time consuming than the exhaustive search.

Let μ_1 be an integer, S be a dictionary, f_b be defined by (5.3.1). First, fix upon an element $b_1 \in GF(2^{\mu_1})$ for which $I(f_{b_1}, S)$ is minimal. By (5.3.1),

$$I_1 = I(f_{b_1}, S) \le T\frac{n}{\mu_1} \cdot 2^{\mu_1} \tag{5.3.4}$$

Define for an element $b \in GF(2^{\mu_1})$ the concatenation $f_{b_1} f_b$, which maps a word $w \in S$ to $f_{b_1}(w) f_b(w)$.

Second, fix upon an element b_2, for which $I(f_{b_1} f_b, S)$ is minimal. By Theorem 5.1.1,

$$I_2 = I(f_{b_1} f_b, S) \le I_1 \frac{n}{\mu_1} \cdot 2^{\mu_1} \tag{5.3.5}$$

Repeat the procedure k times, $k \ge 1$. We get a map $f = f_{b_1} f_{b_2} \dots f_{b_k}$ from E^n to E^μ, $\mu = k \cdot \mu_1$. Inequalities (5.3.4) and (5.3.5) yield

$$I(f, S) \le \left(\frac{n}{\mu_1 2^{\mu_1}}\right)^k \cdot T \tag{5.3.6}$$

The time consumption is $O(kTn \log T \cdot 2^{\mu_1})$. The algorithm may be called greedy: at each step we choose a best possible element. Selecting elements from $GF(2^{\mu_1 k})$ will give a lesser index, but the time consumption will be greater. This algorithm yields Claim 5.3.2.

Claim 5.3.2.

Given a dictionary $S \subseteq E^n$, $|S| = T$, and two integers, μ and k, $\mu 2^\mu < n$, k being a divisor of μ, the greedy algorithm in $O(knT \log T \cdot 2^{\mu/k})$ units of time produces a polynomial map

$$f_b : E^n \to E^\mu,$$

whose index on S is not very great:

$$I(f_b, S) \le T \left(\frac{nk}{\mu}\right)^k \cdot 2^{-\mu}$$

Exhaustive search, taking $O(Tn \log T \cdot 2^\mu)$ units of time, produces a map

$$f : E^n \to E^\mu, \quad I(f, S) \le T\frac{n}{\mu} 2^\mu.$$

The degree of a polynomial is not greater than the degree of the field. It is the explanation of the condition $\mu 2^\mu < n$.

Consider an example. There is a dictionary $S \subseteq E^6$ in Table 5.3.2, $|S| = T = 8$, $\mu = 4$, $\mu_1 = 2$, $k = 2$.

The polynomial $x^2 + x + 1$ is irreducible in $GF(2^2)$. Its root is denoted by Θ. The field $GF(2^2)$ and E^2 are isomorfic:

$$00 \to 0, \quad 01 \to \Theta, \quad 10 \to 1, \quad 11 \to \Theta^2.$$

Each element w of S is represented as

$$w = w_1 w_2 w_3, \quad w_i \in GF(2^2), \quad i = 1, 2, 3.$$

For every $b \in GF(2^2)$ and $w \in S$ there are $f_b(w)$ and $I(f_b, S)$ in the table. The maps f_1, f_Θ and f_{Θ^2} have got the minimal index, which is $5/4$. Take the map f_1. Then the concatenation $f_1 f_\Theta$ has got index $1/2$, $f_1 f_\Theta - 1/4$, $f_1 f_{\Theta^2}$ – zero, i.e., it is an injection. The loading factor of those concatenations is $8/2^4 = 1/2$.

Table 5.3.2.

Polynomial $f_b(w) = w_1 + w_2 b + w_3 b^3$

Words w of a dictionary S	A word w, written as $w = w_1 w_2 w_3$			$f_b(w)$, if b equals			
	w_1	w_2	w_3	0	1	Θ	Θ^2
100100	1	Θ	0	1	Θ^2	Θ	0
100101	1	Θ	Θ	1	1	Θ^2	Θ^2
100111	1	Θ	Θ^2	1	0	0	1
100001	1	0	Θ	1	Θ^2	0	Θ
000100	0	Θ	0	0	Θ	Θ^2	1
000001	0	0	Θ	0	Θ	1	Θ^2
000011	0	0	Θ^2	0	Θ^2	Θ	1
011011	Θ	1	Θ^2	Θ	0	Θ	0
Index $I(f_b, S)$				9/4	5/4	5/4	5/4

Some instances of Claim 5.3.2 are especially interesting.

Corollary 5.3.1.

Let A and B be sets, $T \to \infty$, $\alpha = \frac{|B|}{T} = \text{const}$, $\lim \frac{\log T}{\log |A|} > 0$, $\varepsilon > 0$. There are a constant β and a universal β-hash-set $U_T(A, B, \beta)$ such that

1. 1. $|U_T(A, B, \beta)| \leq |B|$

2. 2. *The running time of functions from $U_T(A, B, \beta)$ is $\mathrm{O}\left(\log |A|\right)$*

3. 3. *The time to precompute for a set S, $S \subseteq A$, $|S| = T$, a map $f \in U_T(A, B, \beta)$, such that $I(f, S) \leq \beta$, is not greater than $T^{1+\varepsilon}$.*

The constant β depends on ε only.

Proof.

Let

$$A = E^n, \quad \mu = \log \frac{T}{\alpha}, \quad \mu_1 = \frac{\mu}{k} \tag{5.3.7}$$

We take liberty to assume μ, μ_1 and k to be integers. By the condition of the Corollary,

$$\log T > \tau n, \quad \tau > 0.$$

We first choose the number k so as to make the precomputing time $O\left(kTn\log T \cdot 2^{\mu/k}\right)$ equal to $O\left(T^{1+\varepsilon}\right)$. We use Claim 5.3.2 for that. Then we can see that the upper bound for the index provided by that Claim is a constant β. We have got a set of maps $U_T(A, B, \beta)$, whose cardinality is

$$2^\mu = \frac{T}{\alpha} = |B|.$$

The loading factor of maps in the set is α. Q.E.D.

Corollary 5.3.2.

Let A and B be sets, T be natural,

$$T \to \infty, \quad \lim \frac{\log T}{\log |A|} > 0, \quad \rho = \frac{\log |B| - \log T}{\log T}, \quad \varepsilon > 0.$$

Then there is a $T^{-\rho+\varepsilon}$ - hash set

$$U_T(A, B, T^{-\rho+\varepsilon}), \quad \left| U_T(A, B, T^{-\rho+\varepsilon}) \right| \leq T^{1+\rho}.$$

It takes $O\left(T(\log T)^C\right)$ units of time to find for $S \subseteq A$, $|S| = T$ a map

$$f \in U_T\left(A, B, T^{-\rho+\varepsilon}\right)$$

such that $I(f, S) \leq T^{-\rho+\varepsilon}$.

Proof.

Let

$$\mu_1 = C\log n, \quad k = \frac{\rho+1}{\mu_1}\log T, \quad n = \log |A|.$$

Then Claim 5.2.2 takes the form of Corollary 5.3.2.

Corollary 5.3.3.

Let A and B be sets,

$$T \leq (\log |A|)^C, \quad C = \text{const}, \quad T \to \infty.$$

Then there is a universal numerator $U_T(A, B)$,

$$|B| \leq (\log |A|)^{\log\log|A|}, \quad |U_T(A, B)| \leq (\log |A|)^{\log\log|A|}.$$

It takes $O\left(T(\log |A|)^2(\log\log |A|)^2\right)$ units of time to precompute for $S \subseteq A$, $|S| = T$, an injective map $f \in U_T(A, B)$. The time to calculate

$$f(w), \quad f \in U_T(A, B), \quad w \in A,$$

is

$$O\left(\log|A| \cdot \log\log|A|\right),$$

given an $O\left(\log|A| \cdot \log\log|A|\right)$ *- sized table of indices.*

Proof.
We let $A = E^n$, $\mu_1 = \log n$, $k = \log n$, $\mu = (\log n)^2$. For a dictionary S, $S \subseteq E^n$, there is a μ-length word $b = b_1 \dots b_k$, which determines a map $f_b : E^n \to E^\mu$ by Claim 5.3.2. Its index meets the inequality

$$I(f_b, S) \leq T \left(\frac{n}{\log n}\right)^{\log n} \cdot 2^{-(\log n)^2}.$$

As it is easy to verify,

$$I(f_b, S) \to 0,$$

if $T \leq n^C$, i.e., f_b is an injection.

The word b is a program of f_b. To calculate $f_b x$, we find $f_{b_1}(x), \dots f_{b_k}(x)$, i.e., k times calculate a polynomial (5.3.1). It takes $n \cdot \log n$ bitoperations, given a table of indices of $GF(2^{\log n})$. Q.E.D.

Corollary 5.3.4. (reducing wordlength)
Let $S \subseteq E^n$, $n \to \infty$, be a dictionary, $\log|S| = \tau n$, $\tau > 0$. Then there are both an integer m,

$$m = 2\log|S| + \beta,$$

β *being a constant, and an injective on S map g, $g : E^n \to E^m$, such that the program length of g is $P(g) = O\left(m \cdot 2^{m/3}\right)$ the running time of g is $O(n)$, the precomputing time of g is*

$$PT(g) = O\left(|S|^{5/3} \cdot n\right)$$

Proof.
Define

$$m = 2\tau n + \beta \tag{5.3.8}$$

the constant β will be fixed later. We assume $\frac{m}{3}$ to be an integer.

By Theorem 5.1.1 and Claim 5.3.2, there are three polynomial maps g_1, g_2, g_3, operating from E^n to $E^{m/3}$ such that the index on S of their concatenation $g = g_1 g_2 g_3$ meets inequality (5.3.9):

$$I(g, S) \leq \frac{|S|}{(2^{m/3})^3} \cdot \left(\frac{n}{m/3}\right)^3 \tag{5.3.9}$$

From (5.3.8) and (5.3.9) we obtain, if β is great enough:

$$I(g, S) \leq \frac{1}{|S|} \tag{5.3.10}$$

It means that the concatenation g is injective on S.

To precompute g we make tables of indices for $GF(2^{m/3})$. It takes

$$O\left(2^{\frac{2m}{3}} \cdot m^2\right) = O\left(|S|^{4/3} \log |S|\right)$$

bitoperations. The choice of polynomials g_1, g_2, g_3 takes $O\left(|S| \cdot 2^{m/3}n\right)$ bitoperations.

So, the precomputing time is as required. It takes $O(n)$ bitoperations to compute any polynomial g_1, g_2, g_3 and their concatenation g. The program of g includes the tables of indices $(O(m \cdot 2^{m/3})$ bits) and those three elements of $GF(2^{m/3})$, which determine g_1, g_2, g_3. Q.E.D.

Polynomial functions easily transform variable - length word into computer addresses.

Suppose a non-zero element $x \in GF(2^{\mu})$ is chosen either at random or by search. Precompute the table TM of multiplication by a given x :

$K = \text{Ind}[X]$;

$TM[0] := 0$;

For $I := 1 \, to \, 2^{\mu} - 1 \, do$

$TM[I] := \text{Exp}\left[\text{Ind}[I]\right] + K \, \text{Mod}\,(2^{\mu} - 1)$.

For each $a \in GF(2^{\mu})$ TM gives $a \cdot x$. Take $w = w_1 w_2 \ldots w_{n/\mu}$. Find $w_1 \cdot x$ (one table read). Add w_2 mod 2. Multiply the result by x, and so on. We obtain $h = f_x(w)$, h being the address of the word w :

$H := 0$;

For $I := 1$ to n/μ do

$H := TM[H]$ for $W[I]$.

Table 5.3.3.

The table TM of multiplication by $x = 134 = 1000\ 0110$

0	134	207	73	93	219	146	20	186	60	117	243	231	97	40	174
183	49	120	254	234	108	37	163	13	139	194	68	80	214	159	25
173	43	98	228	240	118	63	185	23	145	216	94	74	204	133	3
26	156	213	83	71	193	136	14	160	38	111	233	253	123	50	180
153	31	86	208	196	66	11	141	35	165	236	106	126	248	177	55
46	168	225	103	115	245	188	58	148	18	91	221	201	79	6	128
52	178	251	125	105	239	166	32	142	8	65	199	211	85	28	154
131	5	76	202	222	88	17	151	57	191	246	112	100	226	171	45
241	119	62	184	172	42	99	229	75	205	132	2	22	144	217	95
70	192	137	15	27	157	212	82	252	122	51	181	161	39	110	232
92	218	147	21	1	135	206	72	230	96	41	175	187	61	116	242
235	109	36	162	182	48	121	255	81	215	158	24	12	138	195	69
104	238	167	33	53	179	250	124	210	84	29	155	143	9	64	198
223	89	16	150	130	4	77	203	101	227	170	44	56	190	247	113
197	67	10	140	152	30	87	209	127	249	176	54	34	164	237	107
114	244	189	59	47	169	224	102	200	78	7	129	149	19	90	22

An example: $a = 137 = 1000\ 1001$, $a \cdot x = 205 = 1100\ 1101$.

M. Trofimova have experimented with the following dictionaries:

1. twelve months;

2. 19 suffixes of English words;

3. 31 common English words;

4. 35 key words in context (KWIC);

5. 36 selected words of "Pascal";

6. 51 American States;

7. 395 English names from Oxford Advanced Learner's Dictionary;

8. 1756 names from a catalogue of file names, which was taken at random in the Novosibirsk University;

9. 3037 English words from a test for students of the Novosibirsk University.

Table 5.3.4.
Experimental results

Dictionary (its size)	Bitlength of the address	Loading factor	Average index	Standard deviation	Minimal index
1(12)	6	0.187	0.164	0.025	0.0
	5	0.375	0.387	0.049	0.166
2(19)	6	0.297	0.292	0.027	0.0
	5	0.594	0.573	0.051	0.211
3(31)	7	0.242	0.208	0.013	0.0
	6	0.484	0.456	0.042	0.129
4(35)	8	0.137	0.123	0.006	0.0
	7	0.273	0.274	0.017	0.057
5(36)	7	0.281	0.259	0.013	0.0
	6	0.563	0.532	0.025	0.166
6(51)	8	0.199	0.193	0.008	0.039
	7	0.398	0.361	0.014	0.078
7(395)	10	0.386	0.384	0.002	0.253
	5 + 5	0.386	—	—	0.273
8(2756)	10	2.691	2.695	0.002	2.603
	5 + 5	2.691	—	—	2.817
9(3037)	10	2.965	2.962	0.003	2.851
	5 + 5	2.965	—	—	2.894

Note: we use exhaustive search except dictionaries $7-9$ with the bitlength of the address equal to 10.

As it is seen from the Table 5.3.4 the average index is close enough to and sometimes even less than the loading factor. The proximity is even better than the theory predicts. The standard deviation is not very large. It means that nearly all functions of the polynomial family have got nearly equal indices on a given dictionary.

Hash-functions for dictionaries $7-9$ were selected either at random from $GF(2^{10})$ or by two-step greedy algorithm: an element of $GF(2^5)$ first and then another element of $GF(2^5)$. The indices of both functions are quite close to each other. However it takes much less time and space to compute two polynomials f_x, $x \in GF(2^5)$, than one polynomial f_x, $x \in GF(2^{10})$.

The hash-functions f_x, where $x = 27 = 00011011$ or $x = 134 = 10000110$ are good for the dictionaries $1-9$ simultaneously.

5.4 String Matching

We have developed two universal sets of maps, linear (5.2.1) and polynomial (5.3.1). Those maps happen to be useful for randomized string matching.

Let there be a text and a dictionary S of all n_1-length words, which occur in the text, $n_1 \geq 1$. On the other hand, there is a n_1-length word w, which is called a pattern. The problem is to find all occurrences of w in the text. There are many algorithms to solve the problem.

A trivial solution is to check whether $a = w$ for all subwords a of the text, $|a| = n_1$. Another approach was proposed by R. Karp and M. Rabin (1987). Let $g : E^{n_1} \to E^{n_2}$ be a map. It takes a word a to $g(a)$, which is called the fingerprint of a. Compare the fingerprint of a subword a of the text with the fingerprint of the pattern w. If the fingerprints are different, go to the next subword. If they are not, we say that there is a match and compare w with a. A match is false, if $g(a) = g(w)$, whereas $a \neq w$. We would like to avoid false matches. There should be a family of fingerprint maps. Being taken at random, a function should have not very great probability of a false match. As we will see, polynomial maps make a good family. The symbol g_x stands for a polynomial (5.3.1).

Lemma 5.4.1.

Let n_1, n_2 be natural, $S \subseteq E^{n_1}$, $|S| = T$, $r < T$, $w \in E^{n_1}$. Call an element $x \in GF(2^{n_2})$ to be r-bad, if there are in S at least r words $a_1, \ldots, a_r$, for which

$$g_x(a_i) = g_x(w), \quad i = 1, \ldots, r.$$

The quantity of r-bad elements in $GF(2^{n_2})$ is not greater than $T \cdot \frac{n_1}{r n_2}$.

Proof.

Make a T-row, 2^{n_2}-column table, whose (x, a) – element is 0, if $g_x(a) = g_x(w)$, or else it is 1, where $a \in S$, $x \in GF(2^{n_2})$. There are not more than $\frac{n_1}{n_2}$ zeroes in each row of the table, because g_x is a $\lceil \frac{n_1}{n_2} \rceil - 1$ - degree polynomial (5.3.1), $w \neq a$, $a \in S$. There are not more than $T \cdot \frac{n_1}{n_2}$ zeroes in the table. The number of r-bad elements, multiplied by r, is not more than $T \cdot \frac{n_1}{n_2}$. Thus, the quantity of r-bad elements is not greater than $T \cdot \frac{n_1}{rn_2}$, Q.E.D.

If a function g_x produces not more than $\delta \cdot T$ false matches, $\delta > 0$, T being the number of n_1-length words of a text, we will say that g_x δ-solves the string (pattern) matching problem. If $\delta T < 1$, we will say that g_x solves the problem exactly. Take elements of $G(n_1, n_2)$ at random. The probability to get a polynomial, which does not δ-solve the string matching problem, equals the number of δT-bad elements divided by $|G(n_1, n_2)| = 2^{n_2}$.

Claim 5.4.1.

The probability to get a polynomial from $G(n_1, n_2)$, which does not δ-solve the string matching problem for a text with T n_1-length words, is not more than $\frac{n_1}{\delta n_2 2^{n_2}}$. The probability of solving the problem exactly is not more than $\frac{Tn_1}{n_2 2^{n_2}}$.

The proof follows from the Lemma 5.4.1.

Corollary 5.4.1.

For any $\delta > 0$ and $n_2 = \lceil \log n_1 \rceil$ almost any function from $G(n_1, n_2)$ δ-solves the string matching problem, notwithstanding the size of the text. Almost any function from $G(n_1, n_2)$ solves the problem exactly, if $n_2 > (1 - \varepsilon) \log T n_1$, $\varepsilon > 0$.

Karp and Rabin (1987) used another set of fingerprint functions.

5.5 Digital Retrieval

At each step of such a retrieval we take a letter out of a word. Depending on that letter, we make a decision what to do next. We give a formal definition.

Let Δ be a binary tree. Each node x of Δ is labelled by an integer $\nu(x) \neq 0$. A label does not occur twice on any path from the root to a leaf. The maximal value of labels is denoted by $\lambda(\Delta)$. A binary word

$$w = w_1 \dots w_n, \quad |w| = n = \lambda(\Delta),$$

determines a path in Δ by the following rule. Start at the root. Being at a node x, take the letter $w_{\nu(x)} = 0$. Go to the left son of x, if $w_{\nu(x)} = 0$, else go to the right one. Stop at a leaf.

The symbol $\mathrm{son}(x, 0)$ stands for the left son of x, $\mathrm{son}(x, 1)$ for the right one.

A word w determines a path $x^0, \dots x^\omega$ by the rule:

$$x^i = \mathrm{son}(x^{i-1}, w_{\nu(x^{i-1})}) \tag{5.5.1}$$

x^0 is the root, x^ω is a leaf. A labelled tree Δ determines a map of $E^{\lambda(\Delta)}$ to the set of its leaves. This map is defined by (5.5.1) and is denoted by the same letter Δ :

$$\Delta w = x^\omega, \quad |w| = \lambda(\Delta).$$

The letters of w, which are not used to find Δw, make a word $r(w)$ which is called the reminder. In order to obtain $r(w)$, delete $\nu(x^0)$-st,...$\nu(x^\omega)$-st letters of w.

Take a labelled tree Δ and a dictionary $S \subseteq E^{\lambda(\Delta)}$. If the map Δ is a bijection of S on the set of leaves, i.e., Δ has got $|S|$ leaves and $\Delta x_1 \neq \Delta x_2$, $x_1, x_2 \in S$, then Δ is called a digital retrieval tree for S. Such a tree is a program of digital retrieval.

An example of a dictionary is on Table 5.5.1. There is the reminder $r(w)$ next to each word w. A retrieval tree for this dictionary is on Fig. 5.5.1. The labels are encircled.

Table 5.5.1.

A dictionary S and the remainders $r(w)$ of its words w with respect to the retrieval tree Δ of Fig. 5.5.1.

word w					reminder $r(w)$		
0	0	0	0	0	0	0	0
0	1	1	0	1	1	0	1
1	0	0	1	0	0	1	
1	1	0	0	1	1	0	
1	0	1	1	1	1	1	
1	1	1	0	0	0		
1	1	1	1	0	$\emptyset$		
1	1	1	1	1	$\emptyset$		

A leaf x of a tree Δ has got a number Prec x, see Section 2.1. The composition Prec $\circ \Delta$ is a bijection of S on the segment $[0, ... |S| - 1]$ of integers.

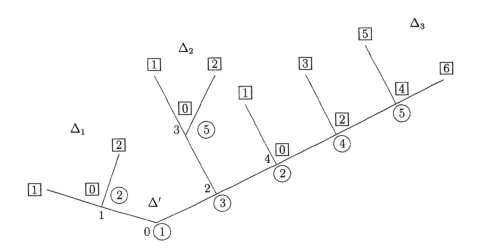

Fig. 5.5.1. A digital retrieval tree Δ for the dictionary of Table 5.5.3. The labels are encircled. The nodes of the basis Δ' are numbered from 0 to 4. The root of Δ_1 is node 1 of Δ'. The root of Δ_2 is node 3 of Δ'. The root of Δ_3 is node 4 of Δ'. The subtrees Δ_1, Δ_2, Δ_3 constitute a partition. Their nodes are numbered within the corresponding trees. The numbers are in squares.

Lemma 5.5.1.

Let Δ be a labelled tree, $\lambda(\Delta)$ be the maximum ot its labels, $|\Delta|$ be the quantity of its nodes, Prec $\circ\Delta$ be the map taking words of $E^{\lambda(\Delta)}$ to integers. A program to calculate Prec $\circ\Delta$ is $O\left(|\Delta|\left(\log(|\Delta| \cdot \lambda(\Delta))\right)\right)$ bits long. The running time is $O\left(\lambda(\Delta) \cdot \log|\Delta|\right)$.

Proof.

The nodes of Δ are numbered from 0 to $|\Delta| - 1$. A node is identified with its number. A leaf x is given the number Prec x too. To calculate Prec $\circ\Delta$ we need the tables of the following four maps.

1. $x \rightarrow \text{son}(x, 0)$.

    The table consists of $|\Delta|$ words, which are $\lceil \log|\Delta| + 1 \rceil$ bits long. A word represents $\emptyset$. There is a number of the left son of x at the place x.

2. $x \rightarrow \mathrm{son}(x, 1)$.
   Analogously.

3. $x \rightarrow \nu(x)$.
   A node x is taken to its label. The table consists of $|\Delta|$ words, each is $\lceil \log \lambda(\Delta) + 1 \rceil$ bits long.

4. $x \rightarrow \mathrm{Prec}\,(x)$, if x is a node, else $x \rightarrow \emptyset$.
   The number of x as a node goes to the number of x as a leaf. The table consists of $|\Delta|$ words, each is $\log |L\Delta(\emptyset)|$ bits long. Here $L(\Delta(\emptyset))$ is the quantity of leaves in Δ, $|L\Delta(\emptyset)| \le |\Delta|$.

Summing the sizes of the tables up, we obtain the bound of Lemma 5.5.1. We have $\lambda(\Delta)$ times operated with $\log |\Delta|$-bits long numbers. Hence, the running time is $\mathrm{O}\,(\lambda(\Delta) \log |\Delta|)$, Q.E.D..

To calculate $\mathrm{Prec} \circ \Delta x$, we use the first, second and third tables and determine the path $x^0, \ldots x^\omega$ by (5.5.1). Sooner or later we meet a leaf x^ω. Its number $\mathrm{Prec}\ x^\omega$ is just $\mathrm{Prec} \circ \Delta x$.

There is a labelled tree on Fig. 5.5.2.

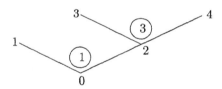

Fig. 5.5.2. *A labelled tree. The labels of nodes are encircled. Leaves are numbered from 0 to 2, from left to right.*

Four tables from the proof of the Lemma are shown on table 5.5.2.

Table 5.5.2.
Four retrieval tables for the tree Fig. 5.5.2.

x	$\mathrm{son}(x,0)$	$\mathrm{son}(x,1)$	$\nu(x)$	$\mathrm{Prec}\,(x)$
0	1	2	1	$\emptyset$
1	$\emptyset$	$\emptyset$	$\emptyset$	0
2	3	4	3	$\emptyset$
3	$\emptyset$	$\emptyset$	$\emptyset$	1
4	$\emptyset$	$\emptyset$	$\emptyset$	2

For $x = 11101$, we have the path 024, $\Delta x = 4$, Prec $\Delta x = 2$.

Take a big enough dictionary S, $\log |S| = cn$. The program of digital retrieval, by Lemma 5.5.1, is $O\left(|S| \log |S|\right)$ bits long.

There is a way to shorten the program of digital retrieval. We make it in two steps.

Let $P = \{\Delta_1, \ldots, \Delta_{|P|}\}$ be a partition of a tree Δ, see Section 2.1. The roots of $\Delta_1, \ldots \Delta_{|P|}$ are the leaves of a tree Δ', which is called the basis of P. The subtrees of P are numbered monotonically, i.e., if $x \in \Delta_i$, $y \in \Delta_l$, $i < l$, then $x < y$. On Fig. 5.5.1 the trees with roots 1,3,4 make a partition. The maximal quantity of nodes in subtrees of P is denoted by $d = d(P, \Delta)$.

Lemma 5.5.2. (two step digital retrieval).

Let Δ be a labelled tree with a partition P, Δ' be the basis of P, $\lambda(\Delta)$ be the maximal label, $|\Delta|$ be the number of its nodes, $d = d(P, \Delta)$.

There is a program to calculate the map Prec $\circ \Delta$. *The program length is*

$$O\left(|\Delta'| \log |\Delta| + |\Delta| \left(\log d + \log \lambda(\Delta)\right)\right)$$

bits. The calculating time is $O\left(\lambda(\Delta) \cdot \log |\Delta|\right)$.

Proof.

The nodes of the basis are numbered from 0 to $|\Delta'| - 1$. To calculate Prec $\circ \Delta$ we need five tables for Δ' and four tables for each subtree Δ_i, $i = 1, \ldots |P|$. The tables for Δ_i are made by Lemma 5.5.1. Their total length is $O\left(|\Delta| \left(\log d + \log |\Delta|\right)\right)$. The first three tables for Δ' are made the same way. The fourth table for Δ' is omitted. Its place is taken by the tables of the following maps.

1) if x is a leaf of Δ', x is mapped to the number j of a subtree Δ_j, for which x is the root. If x is not a leaf of Δ, x is mapped to $\emptyset$. The table of the map consists of $|\Delta'|$ words. Each of them is $\log(|\Delta| + 1)$ bits long.

2) if x is a leaf of Δ', x is mapped to $|L\Delta_1| + \ldots + \left|L\Delta_{j(x)-1}\right|$. The number $j(x)$ is defined above, $|L\Delta_j|$ is the quantity of leaves in Δ_i, $i = 1, \ldots, j(x) - 1$. In other words, x is mapped to the quantity of leaves in subtrees, which are placed to the left of x. The table consists of $|\Delta'|$ words. Their length is $\log |\Delta|$. The total bitlength of all tables meets the bound of the Lemma.

Two-step retrieval of a word w is done the following way.

First, determine by (5.5.1) a path through the basis tree Δ' until a leaf of Δ' is reached. Find the number $j = j(\Delta' w)$ of the subtree, for which $\Delta' w$ is the root. It is done with table 1). Next go to tree Δ_j. Using its tables, go through Δ_j until its leaf Δx is reached. It has got the number

$$\text{Prec} \circ \Delta w = |L\Delta_1| + \ldots + |L\Delta_{j-1}| + \text{Prec} \, ^j(\Delta w)$$

Here $|L\Delta_i|$ is the quantity of leaves of the tree Δ_i, $i = 1, \ldots, j - 1$, Prec $^j(\Delta w)$ is the number, which Δw has got in the tree Δ_j. The sum

$$|L\Delta_1| + \ldots + |L\Delta_{j-1}|$$

is given in the table 2), Prec $^j(\Delta w)$ is in table 1).

The calculation time is $O(n)$ bitoperations.

The retrieval tables for the tree on Fig. 5.5.1 are on Table 5.5.3. Subtrees $\Delta_1, \Delta_2, \Delta_3$ make a partition. Their roots 1,3 and 4 are the leaves of the basis P'. If $w = 11101$, then, by tables for Δ', we get

$$\Delta' w = 4, \quad j = 3, \quad |L\Delta_1| + |L\Delta_2| = 4.$$

Go to the tree Δ_3. The leaf Δw has got in Δ_3 the number 1. Its number in Δ is $|L\Delta_1| + |L\Delta_2| + 1 = 5$.

Table 5.5.3.

Two-step retrieval tables for the tree of Fig. 5.5.1.

First step is Δ', the second step is $\{\Delta_1, \Delta_2, \Delta_3\}$. The nodes are numbered within their subtrees. The label of x is $\nu(x)$, $j(x)$ is the number of subtree, whose root is x, $|L\Delta_1| + \ldots + |L\Delta_{j-1}|$ is the number of leaves of subtrees to the left of x, Prec $^j(x)$ is the number of a leaf x within Δ^j.

Subtree	node	son$(x,0)$	son$(x,1)$	$\nu(x)$	$j(x)$	$|L\Delta_1| + \ldots + |L\Delta_{j-1}|$
Δ'	0	1	2	1	$\emptyset$	$\emptyset$
	1	$\emptyset$	$\emptyset$	2	1	0
	2	3	4	3	$\emptyset$	$\emptyset$
	3	$\emptyset$	$\emptyset$	5	2	2
	4	$\emptyset$	$\emptyset$	2	3	4
						Prec $^j(x)$
Δ_1	0	1	2	2		$\emptyset$
	1	$\emptyset$	$\emptyset$	$\emptyset$		0
	2	$\emptyset$	$\emptyset$	$\emptyset$		1
Δ_2	0	1	2	5		$\emptyset$
	1	$\emptyset$	$\emptyset$	$\emptyset$		0
	2	$\emptyset$	$\emptyset$	$\emptyset$		1
Δ_3	0	1	2	2		0
	1	$\emptyset$	$\emptyset$	$\emptyset$		0
	2	3	4	4		$\emptyset$
	3	$\emptyset$	$\emptyset$	$\emptyset$		1
	4	5	6	5		$\emptyset$
	5	$\emptyset$	$\emptyset$	$\emptyset$		2
	6	$\emptyset$	$\emptyset$	$\emptyset$		3

We will discuss next a way to choose Δ' so as to make the digital retrieval program shorter.

Lemma 5.5.3.

For every dictionary $S \subseteq E^n$, $n \geq 1$, and a number p, $|S| \geq p > 1$, there are both a digital retrieval tree and a partition P such that

    a) the basis Δ' of P has got not more than $2|S|$ nodes

    b) every subtree of P has got not more than $8pn$ nodes.

It takes $O(n^3 |S|)$ bitoperations to build for S such a tree.

Proof.

We will build a retrieval tree Δ for a dictionary S beginning from the root. Every node x of Δ corresponds to a subdictionary $X \subseteq S$. The root corresponds to S. If

$$\max_{1 \leq i \leq n} \max_{\sigma = 0,1} |X(i,\sigma)| = 0,$$

we proclaim the node x to be a leaf of Δ and stop. (Reminder: $X(i,\sigma)$ is the subset of X, whose i-th letter is σ).

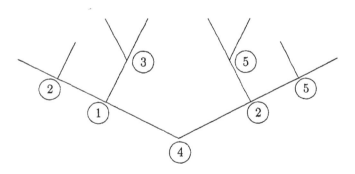

Fig. 5.5.3. *A retrieval tree for the dictionary of table 5.5.1. The labels are encircled.*

If

$$\max_{1 \leq i \leq n} \min_{\sigma = 0,1} |X(i,\sigma)| > 0,$$

we choose any number l, for which

$$\min_{\sigma=0,1} |X(l,\sigma)| = \max_{1 \le i \le n} \min_{\sigma} |X(i,\sigma)|,$$

and label the node x with this very number:

$$\nu(x) = l.$$

Corollary 5.1.2. yields:

$$\min_{\sigma=0,1} |X(\nu(x),0)| \ge \frac{|X|}{4n} \tag{5.5.2}$$

The subset $X(\nu(x),0)$ corresponds to the left son of x, $X(\nu(x),1)$ – to the right one. Go on making the tree until a leaf is reached.

There is a digital retrieval tree made that way for the dictionary of table 5.5.1 on Fig. 5.5.3. The label of the root may be 4 or 5. If it is 4, then its left son may be labelled with 1,3, or 5, etc.

There is a natural order on the set if all partitions of the tree Δ. A partition P^2 majorizes a partition P^1, if for every subtree $\Delta_1 \in P^1$ there is a set of subtrees of P^2, which is a partition of Δ_1. Let M_P be the set of all partitions, all subtrees of which have not less than p leaves. The trivial partition, whose only atom is Δ, belongs to M_P by the condition of Lemma 5.5.3.

Let

$$P = \{\Delta_1, \ldots, \Delta_{|P|}\}$$

be a maximal element of M_P. There are not more than $|P| p$ leaves in Δ, by the definition of M_P. On the other hand, Δ being a retrieval tree for S, the number of its leaves is $|S|$. Thus,

$$|P| \le \frac{|S|}{p}.$$

The number of nodes of the basis equals $2|P| - 1$, which yields the statement a) of the Lemma.

Let x be a root of a subtree Δ_i, $\Delta_i \in P$, X be the subset of words S, which correspond to X, $i = 1, \ldots, |P|$. Suppose that

$$X(\nu(x),0) \ge p, \quad X(\nu(x),1) \ge p.$$

Change the subtree Δ_i in P for two subtrees, whose roots are $son(x,0)$ and $son(x,1)$, correspondingly. The partition obtained belongs to M_p, whereas it majorizes P – a contradiction, since P is maximal. Thus,

$$\min_{\sigma=0,1} |X(\nu(x),\sigma)| < p \tag{5.5.3}$$

From (5.5.2) and (5.5.3) we get

$$|X| \le 4np$$

The tree, whose root is x, has got $|X|$ leaves. It has got

$$2|X| - 1 \leq 8np$$

nodes.

Given a dictionary S, it takes $O(n|S|\log|S|)$ bitoperations to calculate

$$\min_{\sigma=0,1} |S(i,\sigma)|$$

for every i, $1 \leq i \leq n$, since numbers $|S(i,\sigma)|$ are $\log|S|$-bits long at the most. It takes $O(n\log|S|)$ bitoperations to choose a maximal number among those n numbers, see Aho et al. (1983). Label the root of the tree by l,

$$\max_{1 \leq i \leq n} \min_{\sigma=0,1} |S(i,\sigma)| = \min_{\sigma=0,1} |S(l,\sigma)|$$

(Lemma 5.5.3). We have spent $O(n^2|S|)$ bitoperations, since $\log|S| \leq n$. Continue with the sons of the root the same way. It takes

$$O(n^2(|S(l,0)| + |S(l,1)|)) = O(n^2|S|) \tag{5.5.4}$$

bitoperations. So, it takes $O(n^2|S|)$ bitoperations to make a storey of a retrieval tree, and $O(n^3|S|)$ bitoperations to make the entire tree, Q.E.D.

Lemmas 5.5.2 and 5.5.3 being combined, produce a universal numerator. Without loss of generality, we assume $A = E^n$. The loading factor is 1, i.e., there are no holes in the table B, its cardinality equals the cardinality of a dictionary.

Theorem 5.5.1. (two-step digital numeration)

Let A and B be sets, T be natural, $|B| = T$. There is a universal digital numerator $U_T(A, B)$, whose cardinality $|U_T(A, B)|$ meets the upper bound

$$|U_T(A, B)| \leq (\log|A|)^{CT}$$

It takes $O(\log T \cdot \log|A|)$ bitoperations to compute $f(x)$, $f \in U_T(A, B)$, $x \in A$. It takes $O((\log|A|)^3 T)$ bitoperations to precompute for $S \subseteq A$, $|S| = T$, an injective map $f \in U_T(A, B)$.

Proof.
Let for a dictionary $S \subseteq E^n$, $E^n = A$,

$$p = \begin{cases} 2, & |S| < n \\ n, & |S| \geq n. \end{cases} \tag{5.5.5}$$

Then Lemmas 5.5.2 and 5.5.3 yield a program to calculate an injective map Prec $\circ\Delta$: $S^0 \to [0, T-1]$. The time to find such a program is $O((\log|A|)^3 \cdot T)$. The program length by those Lemmas is $O\left(T\log n\right)$ with the parameter p chosen by (5.5.5). Hence, the number of different programs, which upperbounds the cardinality of $U_T(A,B)$, is

$$2^{O\,(T\log n)} = (\log|A|)^{CT}, \quad C = \mathrm{const},$$

Q.E.D.

5.6 Digital Calculation of Boolean Functions

We use digital numerators of Section 5.5 to calculate boolean functions.

Suppose we are given a partial boolean function

$$f : E^n \to \{0,1\}, \quad n \geq 1, \quad \mathrm{dom}\ f = S, \quad |S| = T, \quad T \geq 1.$$

Let $A = E^n$, $|B| = T$. By Theorem 5.5.1, there is a digital map

$$h : E^n \to [0, T-1], \quad h \in U_T(A,B),$$

h is injective on S. Thus, each cluster $h^{-1}(i)$, $0 \leq i \leq T-1$, contains exactly one element of S. Introduce a fully specified (total) boolean function g. On every element of a cluster $h^{-1}(i)$ it takes the value, which is taken by f on the element $h^{-1}(i) \cap S$, $i = 0, 1, \ldots, T-1$:

$$g(h^{-1}(i)) = f(x), \quad x \in h^{-1}(i).$$

To specify g, it is necessary to fix 2^T binary values:

$$g(h^{-1}(0)), \ldots, g(h^{-1}(T-1)).$$

In order to compute g we need, first, a program to compute g, and, second, those values. Thus, the space consumption is $O\left(T\log n\right)$ bits. The functions f and g agree (Section 4.1). Running through all the functions f, $|\mathrm{dom}\ f| = T$, we get a piercing set of all the functions g. We have obtained

Theorem 5.6.1.

For every partial boolean function $f : E^n \to \{0,1\}$, $|\mathrm{dom}\ f| = T$, there is a $O\left(T\log n\right)$-bits length computing program. Its running time is $O(n)$. The time to precompute such a program is $O\left(T \cdot n^3\right)$ bitoperations. There is a digital piercing set $V_T(n)$,

$$|V_T(n)| \leq n^{CT}.$$

Fully defined boolean functions can be calculated digitally.

Theorem 5.6.2.

Let f be a boolean function, which takes the value one on a subset S of E^n :

$$f(x) = 1, \quad x \in S, \quad |S| = T.$$

Then there is a program to calculate f, whose length is

$$T(n - \log T) + CT \log n.$$

If S is nontrivial $\left(\lim \frac{\log T}{n} > 0\right)$, then that program is asymptotically optimal. The time to find such a program is $O\left(Tn^3\right)$ bitoperations.

Proof.
There are $C_{2^n}^T$ boolean functions, which take the value one on some T binary n-length words, $T > 0$. By the Stirling formula, that number equals

$$2^{Tn(n-\log T)(1+o(1))}.$$

Hence, the program lengths of those functions cannot be less than

$$Tn(n - \log T)(1 + o(1)),$$

from which the lower bound of the Theorem follows. To find a program for a function f we proceed by Lemmas 5.5.1 and 5.52. Build a labelled tree Δ and its partition P, the maximal label $\lambda(\Delta) = n$, $p = n$. The basis Δ' of P has got

$$|\Delta'| \leq \frac{2}{n} \cdot T \tag{5.6.1}$$

nodes, the quantity of nodes in subtrees of P is not more than

$$d(\Delta) \leq 8n^2 \tag{5.6.2}$$

(Lemma 5.5.2). There is a map Δ which corresponds to the tree built. The function Prec $\circ \Delta$ maps a word $w \in E^n$ to the number of a leaf of Δ.

The reminder $r(w)$ of w was introduced in Section 5.5. It consists of those letters of w not used to find Δw. There is exactly one word of S, which corresponds to a leaf. The remainder of this word is considered to be the remainder of the leaf. The concatenation (in ascending order) of the remainders of the leafs of a subtree Δ_i is denoted by $r(\Delta_i)$, $i = 1, \ldots |P|$. For its length we have

$$|r(\Delta_i)| \leq n \cdot d(\Delta) \tag{5.6.3}$$

Rewrite (5.6.3), using (5.6.2):
$$r(\Delta_i) = O\left(n^2\right) \tag{5.6.4}$$

The concatenation of all words $r(\Delta_i)$, $i = 1, \ldots, |P|$, taken in ascending order, is denoted by $r(\Delta)$. The length leaf x equals $n - |r(x)|$, by the definition of the reminder.

The Kraft inequality (Claim 2.2.1) takes the form

$$\sum_{x \in L\Delta} 2^{-(n-|r(x)|)} \leq 1 \tag{5.6.5}$$

From (5.6.5) we obtain:

$$|r(x)| = \sum_{x \in L\Delta} r(x) \leq |L\Delta| \, (n - \log |L\Delta|) = T(n - \log T) \qquad (5.6.6)$$

It is just another form of the Shannon Theorem (Claim 2.5.1).

The program to calculate f is composed of four parts. The first is the retrieval program Prec $\circ\Delta$ of the theorem 5.5.1. The second is the word $r(\Delta)$, the concatenation of all remainders. The third is the table of the map

$$x' \to \sum_{y, y < x} |r(y)| \, .$$

Here x is a node of the basis Δ', x' is the number of x within Δ', $x' \leq |\Delta'|$. There are $|\Delta'|$ words in the table, each is

$$\lceil \log T(n - \log T) \rceil = O(n)$$

bits length, (by 5.6.6). The size of the table is $O(T)$ bits, by (5.6.1).

The fourth part of the program consists of $|P|$ subtables.

The i-th of them is the table of the map

$$x_i \to \sum_{y < x, \ y \in L\Delta_i} |r(y)| \, .$$

Here x_i is the number of a leaf $x \in \Delta_i$ within Δ_i, $|x_i| \leq \log |\Delta_i|$. Nonleaves are mapped to $\emptyset$. There are $|\Delta_i|$ words, each $O(\log n)$ bits length in the i-th subtable, by (5.5.2). The total length of all subtables is

$$O \left(\sum_{i=1}^{|P|} |\Delta_i| \cdot \log n \right) = O(T \log n)$$

bits. We can see now, that the total length of the retrieval program (Theorem 5.5.1),and the tables and subtables is $O(T(n - \log T))$. The main contribution to the program length of f is given by the second part, which is the word $r(\Delta)$. Its length is upperbounded by (5.6.6). It is exactly the Claim of the Theorem. When making the program for f, main part of time is devoted to the retrieval program Prec $\circ\Delta$.

Given a word w, $|w| = n$, we calculate $f(w)$ the following way. First, find the leaf $\Delta'w$, $\Delta'w \in \Delta'$, and the leaf $\Delta w \in \Delta$, just as it was done in Lemma 5.5.2. Those letters of w not used to find Δw make the remainder $r(w)$. Let j be the number of the subtree, which $\Delta'w$ belongs to. Go to the third part of the program. Find the quantity of letters of $r(\Delta)$, which correspond to the leaves $y \in \Delta$, $y < \Delta'w$. Omit that many letters from $r(\Delta)$. Go to the fourth part , to its j-th subtable. Find the quantity of letters of $r(\Delta)$, which correspond to the leaves $y \in \Delta_j$, $y < \Delta w$. Omit that many letters more from $r(\Delta)$ and take $|r(w)|$ letters of the word obtained. We have got the word $r(\Delta w)$. Compare $r(w)$ and $r(\Delta w)$. If they coincide, we conclude, that $w \in S$ and $f(x) = 1$. If they are not, $w \notin S$, $f(x) = 0$. Q.E.D.

The following example clarifies the proof. We take the dictionary S of the table 5.5.1. There is a retrieval tree for S on Fig. 5.5.1. The concatenation of reminders $r(\Delta)$ is 0001010110110. The third and the fourth parts of the program are in Table 5.6.1.

Table 5.6.1.

The length of reminders for leaves, preceding a leaf x

The number of a node x, $x \in \Delta'$	0	1	2	3	4							
$\displaystyle\sum_{y,y<x,y\in L\Delta} \|r(y)\|$ – total length of reminders of the leaves, preceding x	$\emptyset$	0	6	6	10							
The number of a node x within Δ_i	Δ_1			Δ_2			Δ_3						
	0	1	2	0	1	2	0	1	2	3	4	5	6
$\displaystyle\sum_{y,y<x,x\in L\Delta_i} \|r(y)\|$ – total length of reminders of the leaves, preceding $x \in \Delta_i$ 0	0	0	3	0	0	2	0	0	2	2	2	2	2

Take, for example, $w = 11101$. We find $\Delta' = 4$, $\Delta w \in \Delta_3$. Go to the table for Δ'. Omit the first ten letters of $r(\Delta)$. The leaf Δw has got number 3 within Δ_3. Go to the subtable for Δ_3. Omit two more letters of $r(\Delta)$. Since $|r(w)| = 1$, take one letter of the word obtained.

We get $r(\Delta w) = 0$. Compare $r(w) = 1$ with $r(\Delta w) = 0$. They are different. Consequently, $f(w) = 0$.

5.7 Lexicographic Retrieval

It is a common method to search for a word in a dictionary. The dictionary S is ordered lexicographically. A word w is compared with the middle word of S. The comparison can take $O(|w|)$ bitoperations. Depending on the result, go to the left or right half of S and continue the same way. The output is the lexicographic number of w in S. The search takes $O(\log|S| \cdot |w|)$ bitoperations in the worst case, which can happen.

There is an improved variant of lexicographic retrieval. It was invented by V. Potapov. The search time is diminished to $O(\log|S| \cdot \log|w|)$ bitoperations.

Denote by w_i the i-th letter of a word w, $1 \le i \le |w|$, by $m(u, w)$ the first mismatch of words u and w, i.e., the number of the first position, at which u and w are different. A binary word w is greater than a binary word u, $(w > u)$, if

$$w_{m(u,v)} = 1, \quad u_{m(u,v)} = 0.$$

Lemma 5.7.1.

Let u, v, w be binary words of equal length, $u < v$, $u < w$, $m(u, v) > m(u, w)$.
Then

$$v < w.$$

Analogously, if

$$u > v, \quad u > w, \quad m(u, v) > m(u, w),$$

Then

$$v > w.$$

Proof.
We will prove the first statement. The letters of u and w are equal up to the first mismatch. By the condition of the Lemma,

$$u_{m(u,w)} = 0, \quad w_{m(u,w)} = 1, \quad u_i = w_i, \quad i < m(u, w).$$

The word v coincides with u up to $m(u, w)$, since $m(u, v) > m(u, w)$. Hence,

$$v_i = u_i, \quad i < m(u, w), \quad v_{m(u,w)} = 0.$$

Consequently, $w > v$, Q.E.D.

By $[u, v]$, $]u, v]$ and $]u, v[$ we denote a closed, half-closed and open intervals of a dictionary S :

$$[u, v] = \{x \in S : \ u \leq x \leq v\}, \quad]u, v] = \{x \in S : \ u < x \leq v\}.$$

The words u and v are upper and lower ends of an interval A :

$$u = l(A), \quad v = u(A),$$

where $A = [u, v]$, or $A =]u, v]$, or $A = [u, v[$.

Denote by $\mathrm{md}(A)$ the middle, $\lceil \frac{|A|}{2} \rceil$-th word of an interval A.

Order a dictionary $S \subseteq E^n$, $n \geq 1$. It takes $O(n|S|\log|S|)$ bitoperations. The dictionary S is an interval itself. So, there are words $l(S)$, $u(S)$, $\mathrm{md}(S)$. Calculate the mismatches $m(l(S), \mathrm{md}(S))$ and $m(u(S), \mathrm{md}(S))$ and put them next to $\mathrm{md}(S)$. The calculation takes $O(n)$ bitoperations.

Consider two half-closed intervals, $[l(S), \mathrm{md}(S)[$ and $]\mathrm{md}(S), u(S)[$. Find their middle words and their mismatches, etc. Such a procedure may be called precomputing. It takes $\lceil \log|S| \rceil$ steps and $O(|S| \cdot n)$ bitoperations. The result of the precomputing is a retrieval program, which is just the ordered dictionary. The ends of intervals and the middle words are marked. The mismatches are also included into the program. Its length is $|S|n(1 + o(1))$.

Given a word $w \in E^n$ and a retrieval program, we want to know the number of w in S. If $w \notin S$, we want to be informed about it.

Potapov's lexicographic retrieval consists of not more than $\lceil \log |S| \rceil$ steps. At the 0-th step we compare one by one the letters of w and $\operatorname{md} S$ up to the mismatch $m(w, \operatorname{md} S)$. We go to the upper half of S, if $w < \operatorname{md} S$, else to the lower one. The mismatch is remembered in order to be used later. If $w = \operatorname{md} S$, then the number of w is $\lceil \frac{|S|}{2} \rceil$, and the search is concluded.

At the i-th stage we know, first, an interval A, which w belongs to, and, second, one of two mismatches, either $m(w, u(A))$ or $m(w, l(A))$. A may or may not include its ends $u(A)$ and $l(A)$. We will determine, which half of A w belongs to. Suppose that

$$m(w, l(A)) = l_i$$

are known, $i = 0, \ldots, \lceil |S| \rceil$. Take the precomputed number $m(\operatorname{md} A, l(A))$. Compare it with $m(w, l(A)) = l_i$. It takes $O(\log n)$ bitoperations, since there are $\lceil \log n \rceil$-digits in those numbers.

There are three options.

1. $m(\operatorname{md} A, l(A)) > m(w, l(A))$.
   Then, by Lemma 5.7.1, $w > \operatorname{md} A$. We conclude, that w belongs to the upper half of A. The roles of A and $l(A)$ will be played by the upper half of A and $\operatorname{md} A$, $l_{i+1} = l_i$.

2. $m(\operatorname{md} A, l(A)) < m(w, l(A))$.
   Analogously, A is changed for its lower half, $l_{i+1} = l_i$.

3. $m(\operatorname{md} A, l(A)) = m(w, l(A)) = l_i$.
   The first $l_i - 1$ letters of $l(A)$, $\operatorname{md} A$ and w coincide, whereas

$$l(A)_{l_i} \neq (\operatorname{md} A)_{l_i}, \quad l(A)_{l_i} \neq w_{l_i}.$$

Hence,

$$(\operatorname{md} A)_{l_i} = w_{l_i}.$$

Take the next, $l_i + 1$-st letters of the words $\operatorname{md} A$ and w, and compare them. Go on comparing up to the $l_{i+1} = m(w, \operatorname{md} A)$-st position. If

$$w_{l_{i+1}} > (\operatorname{md} A)_{l_{i+1}},$$

then the upper half of A will play the role of A, else the lower half will. If $w = \operatorname{md} A$, the search is concluded. It takes not more than

$$O(\log n + l_{i+1} - l_i)$$

bitoperations to choose a half of A which w belongs to.

So, we need

$$\sum_{i=0}^{\lceil \log |S| \rceil} O(\log |S| \cdot \log n + n)$$

bitoperations to make the lexicographic search. We state the result as a Theorem.

Theorem 5.7.1. (V. Potapov)

Let A and B be sets, T be natural, $|B| = T$. There is a universal lexicographic numerator $U_T(A, B)$, whose cardinality meets the upper bound

$$|U_T(A, B)| \leq |A|^{T(1+o(1))}$$

It takes $O\left(\log T \cdot \log\log |A|\right)$ bitoperations to compute $f(x)$, $f \in U_T(A, B)$, $x \in A$.

It takes $O\left(T(\log |A|)^2\right)$ bitoperations to precompute for a dictionary $S \subseteq A$, $|S| = T$, an injective map $f \in U_T(A, B)$.

The theorem is a resume of Potapov's algorithm. We assume $A = E^n$, $|S| = T \to \infty$, $\log |A| \geq \log T$.

A precomputed dictionary is on Table 5.7.1.

Table 5.7.1.
Potapov's Lexicographic Retrieval

			S			
1	0	0	0	0	1	$(-, 3)$
2	0	0	1	0	1	$(3, 1)$
3	0	1	1	1	1	$(2, 1)$
4	1	0	0	0	0	$(1, 3)$
5	1	0	1	0	1	$(1, 2)$
6	1	1	0	0	0	$(2, 3)$
7	1	1	1	0	0	$(2, 4)$
8	1	1	1	0	1	$(5, 4)$
9	1	1	1	1	1	$(4, -6)$

Middle word of S is $\operatorname{md} S = 10101$. Its mismatches are $m(\operatorname{md} S, l(S)) = 1$, $m(\operatorname{md} S, u(S)) = 2$. The lower half is S_0 and consists of 4 words, $\operatorname{md} S_0 = 00101$, its mismatches are $m(\operatorname{md} S_0, l(S)) = 3$, $m(\operatorname{md} S_0, u(S_0)) = 1$.

Take $w = 00000$. We get $m(w, \operatorname{md} S) = 1$, $w < \operatorname{md} S$. Go to S_0. We get

$$u(S_0) = \operatorname{md} S, \quad m(w, u(S_0)) = l_1 = 1 = m(\operatorname{md} S_0, u(S_0)).$$

It is the third option of the algorithm. Compare the letters of w and $\operatorname{md} S_0 = 00101$ from the second position. We find

$$l_2 = m(w, \operatorname{md} S_0) = 3, \quad w_3 < (\operatorname{md} S_0)_3.$$

Go to the upper half of S_0, which is 00001. Compare it with w from the fourth position. We obtain $w \neq S$.

Take $w = 100000$. Then $m(w, \operatorname{md} S) = 3$, go to S_0. We have

$$m(u(S_0), \operatorname{md} S_0) = 1, \quad m(u(S_0), w) = 3.$$

Thus, by Lemma 5.6.1, $w < \mathrm{md}\, S_0$. Go to

$$S_{01} =]00101, 01111, 10000, 10101[, \quad \mathrm{md}\, S_{01} = 10000.$$

Here

$$m(10101, w) = 3, \quad m(10101, \mathrm{md}\, S_{01}) = 3.$$

Comparing $\mathrm{md}\, S_{01}$ with w from the fourth position, we get $w = \mathrm{md}\, S_{01}$, the number of w is 4.

There is a need to make r steps of lexicographic retrieval sometimes, $1 \le r < \lceil \log T \rceil$. Suppose, for simplicity sake, that the cardinality $|S|$ of a dictionary S is a power of two, $|S| = 2$. Subdivide S lexicographically into atoms

$$A_0, \ldots, A_{2^r - 1},$$

where

$$|A_i| = \frac{|S|}{2^r}, \quad i = 0, \ldots, 2^r - 1.$$

The membership function Mem gives each word $w \in S$ a r-length binary number i such that $w \in S_i$:

$$w \in S_{\mathrm{Mem}\, w}.$$

The precedence function Prec is the total size of the atoms preceding $S_{\mathrm{Mem}\, w}$:

$$\mathrm{Prec}\, w = \frac{|S|}{2^r} \cdot \mathrm{Mem}\, w.$$

The program of r–step retrieval is determined by Theorem 5.6.1. It is of $\mathrm{O}\,(2^r \cdot n)$ bits length.

Corollary 5.7.1.
Let S be a dictionary, $|S| = 2^t$, $t \ge 1$, $\{A_0, \ldots A_{2^r - 1}\}$ be a lexicographic partition of S. The program length of Mem *and* Prec *is* $\mathrm{O}\,(2^r \cdot n)$, *their running time is* $\mathrm{O}\,(r \cdot \log n + n)$, *their precomputing time is* $\mathrm{O}\,(|S| \cdot \log |S| \cdot n)$.

5.8 Enumerative Retrieval

It is a variant of the digital retrieval.

Let Δ be a retrieval tree for a source (Section 5.5.). Label its root by 1, its sons by 2, their sons by 3 etc. Then, to find the number of a word w in S, we take the letters of w in order of their appearance. Such a way of retrieval is called enumerative. The words of S are numbered from 0 to $|S| - 1$.

If Δx is the leaf corresponding to a word x, $|L\Delta x|$ is the number of leaves in the subtree, whose root is x, then the precedence function Prec x is defined by (2.1.4):

$$\mathrm{Prec}\, x = x_1\, |L\Delta(0)| + x_2 L\Delta(x_1 0) + \ldots + x_{|\Delta x|} \left| L\Delta(x_1 \ldots x_{|\Delta x| - 1} 0) \right|$$

To compute Prec x we need a table. It consists of $|\Delta|$ words, each is $\log|L\Delta|$ bits long. The number $|L\Delta(x_1, \ldots x_{i-1}0)|$ corresponds to a node $x = x_1 \ldots x_i$, $i \geq 0$. The size of the table is $|\Delta| \log|L\Delta|$. If $E^n = A$, $S \subseteq E^n$, $|S| = T$, then the size of the table is $O\,(T \log T)$. It means that there are T^{CT} functions, whose programs are those tables. To compute a function, we make n additions of $\log T$ - long binary numbers, or $\log|A| \cdot \log T$ bitoperations, by (2.1.4). To prepare a table for a dictionary S, we order it first. It takes $O\,(T \cdot (\log|A|)^2)$ bitoperations. Then we find $L\Delta(0)$. It takes $O\,(T \cdot \log|A|)$ bitoperations. The calculation of $L\Delta(00)$ and $L\Delta(10)$ takes $(L\Delta(00) + L\Delta(10)) \log|A| = O\,(T \log|A|)$ bitoperations, etc. So, to calculate all coefficients of (2.1.4), we need $O\,(T(\log|A|)^2)$ operations. We get

Theorem 5.8.1.

Let A and B be sets, T be natural, $|B| = T$. There is a universal numerator $U_T(A, B)$, whose cardinality meets the upper bound

$$|U_T(A, B)| \leq T^{CT}$$

It takes $O\,(\log|A| \cdot \log T)$ bitoperations to compute $f(x)$, $x \in A$, $f \in U_T(A, B)$. It takes $O\,(T(\log|A|)^2)$ bitoperations to precompute for a dictionary $S \subseteq A$, $|S| = T$, an injective map $f \in U_T(A, B)$.

Denote by $n_S(x_1, \ldots, x_k)$ the quantity of words produced by S and having the prefix $x_1 \ldots x_k$. Rewrite (2.1.4) as

$$\text{Prec } x = \sum_{i=1}^{|x|} x_i n_S(x_1, \ldots, x_{i-1}, 0) \tag{5.8.1}$$

For a k-letter alphabet, $k \geq 2$, (5.8.1) takes the form

$$\text{Prec } x = \sum_{i=1}^{|x|} \sum_{m=1}^{x_i-1} n_S(x_1, \ldots, x_{i-1}, m) \tag{5.8.2}$$

This formula and the following examples are from Cover (1973).

Given an integer m, find a word $x \in S$, whose number Prec x is m, compare first m with $n_S(0)$. If $m > n_S(0)$, let $x_1 = 1$ and change m for $m - n_S(0)$. Else let $x_1 = 0$ and do not change m. Go to determination of x_2, comparing a new value of m ($m - n_S(0)$ or m) with $n_S(x_1, 0)$ and so on.

Next we apply (5.8.2) to some sources.

Enumeration of binary words with a number of units given.

A source S generates n-length binary words, $n > 0$. There are w, $0 < w < n$, units in each word x. There are $n_S(x_1, \ldots x_{k-1}, 0)$ words, whose prefix is $x_1, \ldots, x_{k-1}, 0$. Obviously,

$$n_S(x_1, \ldots, x_{k-1}, 0) = C_{n-k}^{w - \sum_{j=1}^{k-1} x_j},$$

because it is the number of ways to distribute the remaining $w - \sum\limits_{j=1}^{k-1} x_j$ units among the remaining $n - k$ places of a word. Rewrite (5.8.1) as

$$\text{Prec } x = \sum_{k=1}^{|x|} x_k C_{n-k}^{w-\sum\limits_{j=1}^{k-1} x_j} \tag{5.8.3}$$

For example, if $x = 1000101$, $n = 7$, $w = 3$,

$$\text{Prec}(1000101) = C_6^3 + C_2^2 + C_0^1 = 21,$$

$$\text{Prec}(1110000) = C_6^3 + C_5^2 + C_4^1 = 34.$$

The computation by (5.8.3) is performed step by step. Denote $\sum\limits_{j=1}^{k-1} x_j$ by σ.

The number $C_{n-k}^{w-\sigma}$ is known at the k-th step. At the next step divide it by $n - k$ and multiply by either $w - \sigma - 1$, if $x_k = 1$, or $n - k - w + \sigma - 1$, if $x_k = 0$. We get

$$C_{n-k-1}^{w-\sum\limits_{j=1}^{k+1} x_j} .$$

The number $C_{n-k}^{w-\sigma}$ contains n digits, whereas the numbers $n - k$, $w - \sigma - 1$ and $n - k - w + \sigma - 1$ contain $\log n$ digits. It takes $O(n \cdot \log n \cdot \log\log n)$ bitoperations to multiply or divide a n-digit number by a $\log n$-digit one.

We have obtain the folowing

Claim 5.8.1.

To find the number of a binary word x within the set of words with $r_1(x)$ ones, we need $O(n \log n \log\log n)$ *bitoperations per a letter of x.*

MEE - code of Theorem 3.2.1 differs from the number of x within this set by a prefix only. The prefix is determined by the quantity $r_1(x)$ of ones in x. We may tabulate the prefix in a table with n locations.

The redundancy R of the MEE - code on the set of Bernoulli sources is $O\left(\frac{\log n}{n}\right)$, n being the blocklength.

The redundancy and the number of operations are related:

Claim 5.8.2.

If suffices to spend $O\left(\frac{1}{R}\left(\log\frac{1}{R}\right)^{1+\varepsilon}\right)$ *bitoperations to construct a universal R-redundant code, $R \to 0$, ε is an arbitrary positive number.*

Enumeration of permutations.

Let S be the set of all permutations of $\{1, \ldots n\}$, Prec x be the lexicographic number of a permutation $x = (x_1, \ldots, x_n)$. Given x and i, $1 \le i \le n$, r_i stands for the quantity of integers, which, first, are less than x_i and, second, situated to the right of i. Relation (5.8.2) becomes

$$\text{Prec } x = \sum_{i=1}^{n} r_i (n - i)!$$

Enumeration of Markov sources.
Let $x = x_1 \ldots x_n$, $n > 1$ be a binary word, $\nu_{00}, \nu_{01}, \nu_{10}, \nu_{11}$ be the counts of $00, 01, 10, 11$ among

$$x_1 x_2, \ x_2 x_3, \ x_3 x_4, \ldots, x_{n-1} x_n,$$

S_μ be the set of binary words, whose counts of $00, 01, 10, 11$ are $\nu = (\nu_{00}, \nu_{01}, \nu_{10}, \nu_{11})$. Denote by $g(S)$ the quantity of words $x \in S_{nu}$, whose first letter is $0 : \ x_1 = 0$. It is easy to see, interchanging 0 and 1, that

$$|S_\nu| = g(\nu_{01}, \nu_{10}, \nu_{00}, \nu_{11}) + g(\nu_{10}, \nu_{01}, \nu_{11}, \nu_{00}).$$

If $x \in S_\nu$, $x_1 = 0$, then there are $\nu_{10} + 1$ runs of 0 and ν_{01} runs of 1 in x. The runs are alternated. A run of zeroes goes first. If $x_n = 1$, then

$$\nu_{01} = \nu_{10},$$

else

$$\nu_{01} = \nu_{10} + 1.$$

There are

$$\nu_{10} + \nu_{00} + 1$$

zeroes and $\nu_{10} + \nu_{11}$ units in x. If zero runs lengths are

$$r_1, \ldots, r_{\nu_{10}+1},$$

then

$$r_1 + \ldots + r_{\nu_{10}+1} = \nu_{10} + \nu_{00} + 1.$$

That equation has got

$$C_{\nu_{10}+\nu_{00}}^{\nu_{10}}$$

positive solutions. The corresponding number for unit runs is

$$C_{\nu_{10}+\nu_{11}-1}^{\nu_{01}-1}.$$

Multiplying those numbers, we obtain

$$g(\nu) = g(\nu_{01}, \nu_{10}, \nu_{00}, \nu_{11}) = C_{\nu_{00}+\nu_{10}}^{\nu_{10}} C_{\nu_{11}+\nu_{01}-1}^{\nu_{01}-1},$$

if

$$\nu_{01} = \nu_{10} \quad \text{or} \quad \nu_{01} - \nu_{10} + 1.$$

Else
$$g(\nu) = 0.$$

A word
$$x = \bar{x}_k z_{k+1} \ldots z_n, \quad x \in S_\nu,$$

is a continuation of a word
$$x_1 \ldots x_{k-1} \bar{x}_k,$$

i.e.,
$$x_1 \ldots \bar{x}_k z_{k+1} \ldots z_n \in S_\nu$$

iff
$$\nu(\bar{x}_k z_{k+1} \ldots z_n) = \nu - \nu(x_1, \ldots, \bar{x}_k).$$

Hence, the number Prec x of a word x in S_{nu} is
$$\text{Prec } x = \sum_{k=1}^{n} x_k g(\nu - \nu(x_1 \ldots x_{k-1} \bar{x}_k))$$

If
$$x = (0, 1, 1, 0, 0, 1, 1, 0),$$

then
$$\nu(x) = (\nu_{01}, \nu_{10}, \nu_{00}, \nu_{11}) = (2, 2, 1, 2).$$
$$\text{Prec } x = x_2 g(2, 2, 0, 2) + x_3 g(1, 1, 1, 2) + x_6 g(1, 1, -1, 1) + x_7 g(0, 0, 0, 1) =$$
$$= C_2^2 C_3^1 + C_2^1 C_2^0 + 0 + 0 = 5.$$

To calculate the number Prec x by (5.8.1) we need to know all numbers $n_S(x_1, \ldots x_{i-1}, 0)$. There is no such a need for finite state sources (Section 1.1). Denote by ν_i^t the number of t-length words, which are produced by a source S starting from a state i, $i = 1, \ldots, k$, $t = 0, 1, \ldots$. The source S has k states. The number ν_i^0 is 1, if the i-th state is final, else $\nu_i^0 = 0$. Introduce a vector
$$N^t = (\nu_1^t, \ldots, \nu_k^t).$$

We have
$$N^t = S N^{t-1}, \quad N^t = S^t N^0,$$

where S is the matrix of the source S. For a k-dimensional vector y $\prod(a, y)$ stands for the a-st coordinate of y, $\prod(\emptyset, y) = 0$. Denote by $S(x_1, \ldots, x_i)$ the state, which the source will be in after producing a word $x_1, \ldots, x_i$, $i = 0, 1, \ldots, n$. $S_l(x_1, \ldots, x_i)$ stands for the left brother of the state $S(x_1, \ldots, x_i)$. If a node is either a left, or the only son, or has no father at all, then it has not got the left brother. Now formula (5.8.1) may be rewritten as

$$\text{Prec } x = \sum x_i \prod \left(S_l(x_1, \ldots, x_i), N^{n-i} \right) \tag{5.8.4}$$

Claim 5.8.3.

Let a k-state source S generates n-length binary words, $k \geq 1$, $n \geq 1$. There is a $Ck(n + \log k)$ bits long program, which gives each word $x \in S$ its lexicographic number.

Proof.

Make a program by formula (5.8.4). Prepare two tables, each consisting of $k \lceil \log k \rceil$-bits words. The i-th word of the first table is the number of the state, where the source goes to from the i-th one after generating 0, the i-th word of the second one – after generating 1. The vector N^0 is given. For a word $x = x_1 \ldots x_n$ determine a trajectory

$$S(x_1), \ldots, S(x_1, \ldots, x_n)$$

and a left brothers trajectory

$$S_l(x_1), \ldots, S_l(x_1, \ldots, x_n).$$

To determine $S_l(x_1, \ldots, x_i)$, take $S(x_1, \ldots, x_{i-1})$ and find both $S(x_1, \ldots, x_{i-1}, 0)$ and $S(x_1, \ldots, x_{i-1}, 1)$ by those tables. If

$$S(x_1, \ldots, x_{i-1}, 0) \neq S(x_1, \ldots, x_{i-1}, 1)$$

then

$$S_l(x_1, \ldots, x_i) = S(x_1, \ldots, x_{i-1}, 0),$$

else

$$S_l(x_1, \ldots, x_i) = \emptyset.$$

We need $n\lceil \log(k+1) \rceil$ bits to record both trajectories. The symbol $\emptyset$ can appear.

Make a subroutine to find N^t from N^{t-1}. Each coordinate of N^t is obtained by adding two coordinates of N^{t-1}. Thus,

$$\nu_i^t \leq 2^t.$$

Hence, N^t is represented by k binary words, each being n bits long. Proceed by (5.8.4). Take x_0 and $S_l(x_1, \ldots, x_n)$. Compute N^1 and put it at the place of N^0, etc. The memory consumption meets the Claim. Q.E.D.

Two-step enumeration is analogous to two-step digital retrieval.

Lemma 5.8.1.

Let Δ be a uniform tree, h be its height, $|\Delta(\emptyset)|$ be the quantity of its nodes, $L(\Delta(\emptyset))$ be the quantity of its leaves, $\text{Prec}\,(x)$ be the quantity of its leaves to the left of a node x, $0 < p < R < h$. Then there is a program to compute the map $x \rightarrow \text{Prec}\,x$, $|x| = R$, whose length is

$$(\log |\Delta(\emptyset)| + 1)2^p + 2^R \left(\log \frac{|L\Delta(\emptyset)|}{2^p} + R - p \right)$$

bits.

Proof.

Take relation (2.1.4). Create two tables. Put an integer

$$x_1 |L\Delta(0)| + \ldots + x_p |L\Delta(x_1 \ldots x_{p-1}0)|$$

at val $(x_1 \ldots x_p)$-st place of the first table. Its bitlength is $2^p \cdot (\log |\Delta(\emptyset)| + 1)$. Put an integer

$$x_{p+1} |L\Delta(x_1 \ldots x_p0)| + \ldots + x_R |L\Delta(x_1, \ldots x_{R-1}0)|$$

at val $(x_1 \ldots x_R)$-st place of the second table. The table being uniform, that integer is not greater than

$$2^{R-p} \cdot \frac{|L\Delta(\emptyset)|}{2^R}$$

Hence, the table length is

$$\log \frac{|L\Delta(\emptyset)|}{2^p} + R - p.$$

Total length of both tables meets the bound of the Lemma. Given the tables, it is straightforward to find Prec x by 2.1.4. Q.E.D.

There is another variant of enumerative coding. First, calculate the indicator of a source S, i.e., a 2^n-length word, whose val x-st coordinate is 1, if $x \in S$.

Second, rank all binary 2^n-length words with $|S|$ units from 0 to $C_{2^n}^{|S|-1}$. S is given its indicator's rank, which is $\log C_{2^n}^{|S|}$ bits long (in binary). If S is asymptotically nontrivial $\left(\frac{\log|S|}{n}\right) > 0$, then by Stirling formula the length of the rank is asymptotically $|S| (n - \log |S|)$. The rank may be considered an enumeration program. To find the number of x, $x \in S$, we have to count the units of the indicator to the left of val x-st place. The running time is $O(2^n)$. That program is shorter than the program of either lexicographic or digital retrieval.

NOTES

Enumerative encoding is due to T. M. Cover (1973). It is well suited for the universal compression of Markov and Bernoulli sources.

Usual variants of lexicographic and digital retrieval are in Knuth (1973).

A faster lexicographic search (Section 5.7) was invented by V. Potapov.

There are two ideas behind the improved two-step digital retrieval of this book. First, start the search from a digit splitting the dictionary into nearly equal parts. Second, numerate the nodes within their subtrees (beginning from a level). Incorporate into such a retrieval an idea of O. B. Lupanov: to determine whether a word w belongs to a dictionary S, reach first a leaf of a retrieval tree. Then employ the letters not used. The two-step retrieval along with that idea yields an optimal program to calculate boolean functions.

The experiments with polynomial hash-functions were done by M. Trofimova.

CHAPTER 6

Optimal Numerator

Exhaustive search gives asymptotically best numerators. Elementary numerators (digital, lexicographic, Galois) are easy to construct. On the other hand, their sizes are not optimal. In this chapter we will combine elementary numerators with numerators produced by exhaustive search over small-sized dictionaries. We will aim at three targets. First, the size of the numerator obtained should be small. Second, the numerator should be simple enough to build. Third, its component functions should be fast to calculate. All those targets can be hit nearly simultaneously. It means that each dictionary S is given a perfect hash-function of minimal programlength, which is $\log_2 e$ bits per word. The time to find (precompute) such a function is polynomial in the size of the dictionary. The running time of the function is linear in the wordlength. It is one of the main results of the book. There is an entropy doubling effect. If words of a dictionary S are given codes, whose length exceeds the doubled Hartley entropy of S, then the programlength of a perfect hashing function is linear in $\log |S|$. If that length is less than the dobled entropy, then the programlength is linear in $|S|$.

For nearly any linear code correcting arbitrary additive noise the Varshamov-Gilbert bound is tight.

The algorithm producing perfect hash-functions yields good piercing sets, independent sets and threshold formulas as well.

Using universal numerators, we obtain tables for Lipschitz functions that are optimal by its length and computing time at once.

6.1 Lexicographic-Polynomial Partitions

We want to construct a function f injective on a given dictionary S. We begin with making a special partition

$$A = \{A_1, \ldots, A_{|A|}\}, \quad |A| \geq 1,$$

of S. We denote by Mem $_A w$ the number of the atom of A, which a word w belongs to:

$$w \in A_{\text{Mem }_A w}.$$

The map Mem $_A$ is called the membership function of A.

Every word w is one-to-one represented in A by a shorter word Rep $_A w$. Its length is n_A : $|\text{Rep }_A w| = n_A$. The map Rep $_A$ is called the representation function of A. The values of the representation function within an atom differ from each other.

We denote by $s(A)$ the maximal size of atoms of A :

$$s(A) = \max_{1 \leq i \leq |A|} |A_i|$$

By $\gamma(A)$ we denote the quantity of different sizes of atoms of A. The precedence function Prec maps a word w to the combined size of the preceding atoms:

$$\text{Prec } w = \sum_{l < \text{Mem } w} |A_l|$$

Our aim is to find a partition A such that, on one hand, its atoms are small enough. On the other hand, its membership, representation and precedence functions are simple enough. If $A = \{S\}$, then Mem is trivial, whereas the only atom of A is large. If A consists of one-word atoms, then Mem is a rather complicated function. We will look for a good partition A somewhere in between those extremes.

Lemma 6.1.1.

Let $S \subseteq E^n$. There is a partition A of S such that

1. *The program complexities of* Mem $_A$, Rep $_A$, Prec $_A$ *are*

$$\log n (1 + \text{o}\,(1)) + \text{o}\,(|S|).$$

2. *Their precomputing times are*

$$\text{O}\,(|S|^3\, n^2).$$

3. *Their running times are*

$$\text{O}\,(n \log n)$$

4. *Every atom of A contains* $\text{O}\,(l_2(n) \cdot l_3(n))$ *words:*

$$s(A) = \text{O}\,(l_2(n) \cdot l_3(n)), \quad \gamma(A) = 1.$$

5. *Every word $w \in S$ is one-to-one represented in A by a word* Rep $_A w$,

$$|\text{Rep }_A w| = \text{O}\,(l_3(n)).$$

Proof.
All parameters are assumed to be integers for simplicity sake. We will make a partition required step by step starting with the trivial partition

$$B = \{B_1\}, \quad B_1 = S, \quad \text{Rep }_B w = w$$

Suppose

$$X = \{X_1, \ldots, X_{|X|}\}$$

is a partition of S, the number of atoms of X is a power of two:

$$|X| = 2^{k(X)}.$$

Every atom of X contains

$$s(X) = |S| \cdot 2^{-k(X)} \tag{6.1.1}$$

words, $\gamma(X) = 1$. The words of S are one-to-one represented in X by $n(X)$-length words. We will reduce X to a new partition Y. The reduction is carried out by a map $X \to Y$. Being fed with both the representative Rep $_X w$ of w in X and Mem $_X w$, $X \to Y$ produces Rep $_Y w$ and Mem $_Y w$. Symbols Pr$(X \to Y)$ and PT$(X \to Y)$ stand for the program length and precomputing time of the map $X \to Y$.

There are three options to go from X to Y. The first one is to choose for each atom X_i, $i = 1, \ldots, |X|$, an injective on it polynomial f_b, where

$$n(Y) = |b| = 2 \log s(X) + \log n(X) \tag{6.1.2}$$

A polynomial f_b is defined by (5.3.1). The choice is done by exhaustive search (Claim 5.3.1). The polynomial f_b being selected, the representatives of words in the atom X_i are changed. Namely, we define for any word w

$$\text{Rep } _Y w = f_b(\text{Rep } _X w) \tag{6.1.3}$$

Atoms of X become atoms of Y, by definition. All the difference between X and Y is made by the representatives of words of S in those partitions:

$$n(X) = |\text{Rep } _X w| > n(Y) = |\text{Rep } _Y w|.$$

The program of $X \to Y$ is the concatenation of all words b chosen for all atoms:

$$\Pr(X \to Y) = |X| \cdot n(Y) = \frac{|S| \cdot n(Y)}{s(X)} \tag{6.1.4}$$

The precomputing time of $X \to Y$ is given by Claim 5.3.1:

$$\text{PT}(X \to Y) = \text{O}\left(|X| \cdot (s(X))^3 \cdot (n(X))^3\right) = \text{O}\left(|S| \cdot S^2(X) \cdot n^3(X)\right) \tag{6.1.5}$$

The second option to go from a given partition X to a partition Y may be exercised if

$$s(A) \leq n_A^C,$$

C being a constant.

An injective polynomial f_b is chosen for each atom X_i again. This time the selection is done not by exhaustive search, but by Corollary 5.3.3 instead. Such a selection is less time consuming. The representatives are changed by (6.1.3).

Corollary 5.3.3 yields:

$$|\text{Rep }_Y w| = n(Y) = \text{O}\left(\log^2 n(X)\right)$$

$$\Pr\left(X \to Y\right) = \text{O}\left(|X| \cdot n(Y)\right) = \frac{|S|}{s(X)} \cdot n(Y)$$

$$\text{PT}\left(X \to Y\right) = \text{O}\left(|X| \cdot s(X)n^2(X)\log^2 n(X)\right) = \text{O}\left(|S| \cdot s(X)n^2(X)\log^2 n(X)\right)$$
$$\tag{6.1.6}$$

The third option to go from X to Y is to define

$$k(Y) = \log|S| - l_1(n(X)) - l_2(n(X)) \tag{6.1.7}$$

and to make $k(Y) - k(X)$ steps of lexicographic retrieval on each atom of X. The partition Y consists, by definition, of

$$2^{k(Y)} = |Y| = \frac{|S|}{n(X)\log n(X)} \tag{6.1.8}$$

atoms obtained. The representatives of words remain unchanged:

$$\text{Rep }_Y w = \text{Rep }_X w, \quad n(X) = n(Y).$$

The program $X \to Y$ is the concatenation of $|X|$ program for atoms of X. We obtain from Corollary 5.7.1:

$$\Pr\left(X \to Y\right) = \text{O}\left(|Y| \cdot n(X)\right) = \text{O}\left(\frac{|S|}{s(Y)}n(X)\right)$$

$$\text{PT}\left(X \to Y\right) = \text{O}\left(n^2(X) \cdot |Y|\right) \tag{6.1.9}$$

The lexicographic retrieval produces the precedence function automatically. That function remains unchanged for the first and second options when going from X to Y.

Start with the trivial partition B. Choose one of three options and transform B to a partition C. Then transform C to D etc. At last we will obtain the partition A required.

Five different cases present themselves depending on the relation between the size $|S|$ and the wordlength n. For each case the way from the first B to the last partition A is illustrated by a diagram. The names of partitions are encircled on Fig.6.1.1-6.1.5.

The option chosen to go from a partition to the next one is put above the arrow.

Let X be a partition, Y be the next one. Having got $n(X)$ and $s(X)$, we find $n(Y)$ by (6.1.2) (first option), (6.1.6) (second) or let $n(Y) = n(X)$ (third). For the size of atoms we get either $s(Y) = s(X)$ (first and second options), or find $s(Y)$ by (6.1.8):

$$s(Y) = n(X)\log n(X) \quad \text{(third option)}.$$

When we know $n(Y)$ and $S(Y)$, we will calculate the programlength $\Pr(X \to Y)$ and precomputing time $\mathrm{PT}(X \to Y)$ by (6.1.4)-(6.1.6) or (6.1.9).

Summing all the expenses on the way from the first partition B to the partition required A, we will get the estimations $1, 2, 4$ and 5 of the Lemma.

To calculate Mem_A or Rep_A we have to make either several calculations of Galois polynomial or lexicographic search. The running time of both does not exceed $n \cdot \log n$. That gives the third statement of the Lemma.

Next we go to those five cases.

1. First case.

$$|S| \geq n(\log n)^2.$$

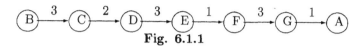

Fig. 6.1.1

Here we have

$$n(B) = n(C) = n,$$

$$s(B) = |S|, \quad s(C) = s(D) = \mathrm{O}(n \log n), \quad \text{see} \quad (6.1.8)$$

$$n(D) = n(E) = \mathrm{O}(\log^2 n), \quad \text{see} \quad (6.1.6)$$

$$s(E) = s(F) = \mathrm{O}(l_1^2(n)l_2(n)), \quad \text{see} \quad (6.1.8)$$

$$n(F) = n(G) = \mathrm{O}(l_2(n)), \quad \text{see} \quad (6.1.2)$$

$$s(G) = s(A) = \mathrm{O}(l_2(n)l_3(n)), \quad \text{see} \quad (6.1.8)$$

$$n(A) = \mathrm{O}(l_3(n)), \quad \text{see} \quad (6.1.2)$$

$\Pr(\mathrm{Mem}_A)$ is the sum of $\Pr(B \to C)$, etc., $\Pr(G \to A)$. $\Pr(E \to F)$ and $\Pr(G \to A)$ are given by (6.1.4), $\Pr(C \to D)$ by (6.1.6), $\Pr(B \to C)$, $\Pr(D \to E)$ and $\Pr(F \to G)$ by (6.1.9). The dominant addend is

$$\Pr(B \to C) = \mathrm{O}\left(\frac{|S|}{\log n}\right) = \mathrm{o}(|S|).$$

The dominant addend among $\mathrm{PT}(B \to C)$, etc., $\mathrm{PT}(G \to A)$ is

$$\mathrm{PT}(C \to D) = |S| \cdot n^3 \log^3 n, \quad \text{by (6.1.6)}$$

So, $\mathrm{PT}(\mathrm{Prec}_A) = \mathrm{PT}(\mathrm{Mem}_A) = \mathrm{O}(|S|^3 \cdot n^2)$

2. Second case.

$$n(\log n)^2 \geq |S| \geq (\log n)^3$$

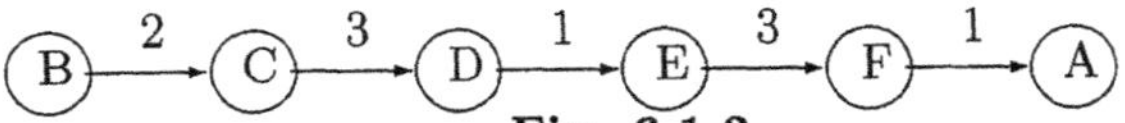

$$\text{(B)} \xrightarrow{2} \text{(C)} \xrightarrow{3} \text{(D)} \xrightarrow{1} \text{(E)} \xrightarrow{3} \text{(F)} \xrightarrow{1} \text{(A)}$$

Fig. 6.1.2

We deal with this case the same way.

$$n(C) = n(D) = \log^2 n, \quad s(D) = s(E) = O\left(\log^2 n \cdot l_2(n)\right),$$

$$n(E) = n(F) = O\left(l_2(n)\right), \quad s(F) = s(A) = O\left(l_2(n) \cdot l_3(n)\right), \quad n = O\left(l_3(n)\right).$$

$\Pr(\text{Mem})$ equals asymptotically

$$\Pr(B \to C) = o\left(|S|\right),$$

$\mathrm{PT}(\text{Mem})$ equals asymptotically

$$\mathrm{PT}(B \to C) = O\left(|S|^2 n^2 \log^2 n\right)$$

3. Third case.

$$l_2^2(n) \le |S| \le (\log n)^3$$

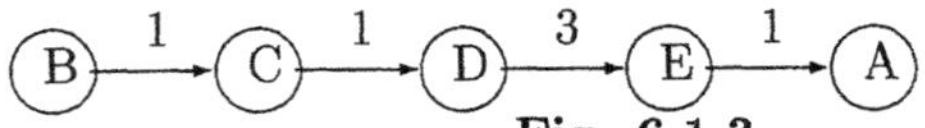

$$\text{(B)} \xrightarrow{1} \text{(C)} \xrightarrow{1} \text{(D)} \xrightarrow{3} \text{(E)} \xrightarrow{1} \text{(A)}$$

Fig. 6.1.3

$$n(C) = \log n(1 + o(1)), \quad n(D) = O\left(l_2(n)\right),$$

$$|S| = s(C) = s(D), \quad s(E) = s(A) = O\left(l_2(n) \cdot l_3(n)\right), \quad n(A) = O\left(l_3(n)\right),$$

$$|\text{Mem }_A| \sim \Pr(B \to C) = \log n(1 + o(1)).$$

$$\mathrm{PT}(\text{Mem }_A) \sim \mathrm{PT}(B \to C) = O\left(|S|^3 \cdot n^3\right)$$

4. Fourth case.

$$l_3^2(n) \le |S| \le l_2^2(n)$$

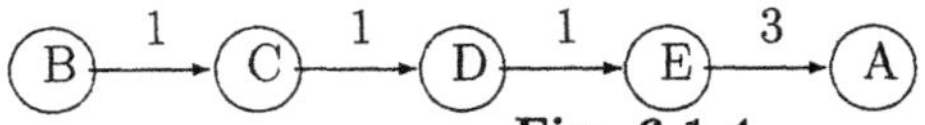

$$\text{(B)} \xrightarrow{1} \text{(C)} \xrightarrow{1} \text{(D)} \xrightarrow{1} \text{(E)} \xrightarrow{3} \text{(A)}$$

Fig. 6.1.4

$$n(C) = \log n(1 + o(1)), \quad n(D) = O\left(l_2(n)\right),$$

$$n(E) = n(A) = O\left(l_3(n)\right), \quad |S| = s(B) = s(C) = s(D) = s(E),$$

$$s(A) = O\left(l_3(n) \cdot l_4(n)\right),$$

$$|\Pr(\text{Mem }_A)| = \log n(1 + O(1)),$$

$$\mathrm{PT}(\text{Mem }_A) - O\left(|S|^3 n^3\right)$$

5. Fifth case.

$$|S| \le l_3^2(n)$$

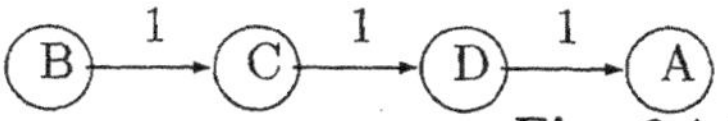

Fig. 6.1.5

$$n = n(B), \quad n(C) = \log n(1 + o(1)), \quad n(D) = O(l_2(n)), \quad n(A) = O(l_3(n)),$$
$$|\mathrm{Pr}(\mathrm{Mem}_A)| = \log n(1 + o(1)),$$
$$\mathrm{PT}(\mathrm{Mem}_A) = O(|S|^3 \cdot n^3), \quad \text{by (6.1.5)}$$

conditions 1., 2., 4. and 5. of the Lemma are met.

Given the program Mem $_A$ and a word w, we can find the atom Mem $_A w$, which w belongs to, and the representative Rep $_A w$. We do it passing from the partition B to the partition A via intermediate partitions. The third and second options take $O(n \log n)$ bitoperations, whereas the first one takes $O(n)$ operations. A table of indices of $GF(2^{\log n})$ is used for computations of Galois polynomials. The size of the table is $O(n \log n)$, the precomputing time is $O(n^2 \log n)$.

We illustrate the idea of the Lemma by an example. Let $S \subseteq E^4$, $S = \{0, 1, 2, 6, 7, 8\}$. The number val w stands for a 4-dimensional vector w. Go from the trivial partition $B = \{S\}$ to a partition A by diagram 6.1.6

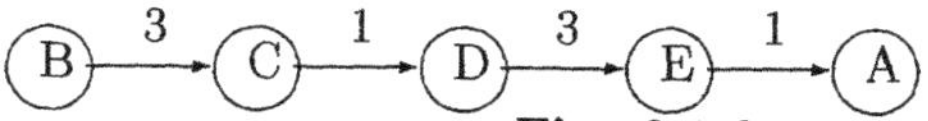

Fig. 6.1.6

One-step lexicographic partition with the middle word 2 transforms B to

$$C = \{C_0, C_1\}, \quad C_0 = \{0, 1, 2\}, \quad C_1 = \{6, 7, 8\}.$$

Each word is represented by itself in B and C,

$$\mathrm{Rep}\ _B w = \mathrm{Rep}\ _C w = w.$$

Select Galois polynomials: f_1 for C_1 and f_3 for C_2. We define

$$\mathrm{Rep}\ _D w = \begin{cases} f_1(w), & \text{if } w \in C_1 \\ f_3(w), & \text{if } w \in C_2 \end{cases}$$

Functions f_1 and f_2 are given by table 5.3.1. Change w for Rep $_D w$. We get

$$D = \{D_0, D_1\}, \quad D_0 = \{0, 1, 2\}, \quad D_1 = \{0, 3, 2\}.$$

Make one-step lexicographic partition with the middle word 1 on D_0 and the middle word 2 on D_1. We get a partition

$$E = \{E_0, E_1, E_2, E_3\}, \quad E_0 = \{0, 1\}, \quad E_1 = \{2\}, \quad E_2 = \{0, 2\}, \quad E_3 = \{3\}.$$

Choose functions from $G(2,1)$ for atoms of E : f_1 for E_0, f_0 for E_1, E_2, E_3 where

$$w = (w_1, w_2), \quad f_0(w) = w_1, \quad f_1(w) = w_1 + w_2.$$

We let

$$\text{Rep}\,_A w = \begin{cases} f_1 w, & w \in E_1 \\ f_0 w, & w \in E_2, E_3, E_4 \end{cases}.$$

Change w for $\text{Rep}\,_A w$. It yields

$$A = \{A_0, A_1, A_2, A_3\}, \quad A_0 = \{0, 1\}, \quad A_1 = \{1\},$$

$$A_2 = \{0, 1\}, \quad A_3 = \{1\}.$$

Table 6.1.1 shows the transition from B to A. There are both the decimal number of the atom Mem $_X w$, which w belongs to in a partition X and the representative Rep $_X w$ (binary)

Table 6.1.1.

Transition from B to A.

Mem $_X w$ is above (decimal), Rep $_X w$ is below (binary).

word w (decimal)	0	1	2	6	7	8
B	0	0	0	0	0	0
	0000	0001	0010	0110	0111	1000
C	0	0	0	1	1	1
	0000	0001	0010	0110	0111	1000
D	0	0	0	1	1	1
	00	01	10	00	11	10
E	0	0	1	2	3	2
	00	01	10	00	11	10
A	0	0	1	2	3	2
	0	1	1	0	1	1

The precedence function for E and A is Prec,

$$\text{Prec}\,(0) = \text{Prec}\,(1) = 0, \quad \text{Prec}\,(2) = 2,$$

$$\text{Prec}\,(6) = \text{Prec}\,(8) = 3, \quad \text{Prec}\,(7) = 5.$$

6.2 Multilevel Galois Partitions

Lemma 6.2.1. (atom splitting)
Let S be a set of n-length words, $n \to \infty$, $A = \{A_1, \ldots, A_{|A|}\}$ be a partition of S, a and b be parameters, $a > b$, $b \to \infty$,

$$s(A) = a + o(a).$$

Then there is a partition B of S, $B = \{B_1, \ldots, B_{|B|}\}$, such that

1. $S(B) \leq b + 1$

2. $|B| \leq \frac{2|S|}{b}(1 + o(1))$

The map $A \to B$ transforming A to B, has the following characteristics:

$$\Pr(A \to B) = O\left(\frac{|S|}{b}(n + l_2(|S|))\right),$$

$$\mathrm{PT}(A \to B) = O(|S| \cdot 2^n \cdot n),$$

$$T(\mathrm{Mem}\ _B) = T(\mathrm{Mem}\ _A) + O(n + \log |S|).$$

Proof.
First, we need tables of indices for Galois fields $GF(2^r)$,

$$r \leq \log \frac{a}{b} + 1.$$

The size of those tables is $O\left(\frac{a}{b} \cdot n\right)$, the precomputing time, by Section 5.2, is $O\left(\frac{a^2}{b^2} \cdot n^3\right)$.

Given an atom A_i, $i = 1, \ldots, |A|$, let

$$m_i = \log \frac{|A_i|}{b} \tag{6.2.1}$$

By Claim 5.2.1, there is a linear function $\varphi : E^n \to E^{m_i}$ such that

$$I(\varphi_i, A_i) \leq b. \tag{6.2.2}$$

The characteristics of φ_i are:

$$\Pr(\varphi_i) = O(n), \quad \mathrm{PT}(\varphi_i) = O(2^n |A_i| \cdot n),$$

$$T(\varphi_i) = O(n) \tag{6.2.3}$$

when computing φ_i, tables of indices are used.

An atom A_i is subdivided by φ_i into 2^{m_i} clusters C_{ij}, $j = 1, \ldots, 2^{m_i}$. The words of a cluster are given the same address by φ_i. If $w \in A_i$, then

$$w \in C_{i\varphi_i(w)}.$$

Lemma 2.4.3 and (6.2.2) yield an upper bound on the size of atoms:

$$|C_{ij}| \le \sqrt{ab(1 + o(1))} \tag{6.2.4}$$

By the definition of index,

$$I(\varphi_i, A_i) = \frac{1}{|A_i|} \sum |C_{ij}|^2 - 1 \tag{6.2.5}$$

There are two groups of clusters. The first group includes the clusters, whose sizes are not greater than b; the second group includes all others. Denote by l the quantity of clusters in the first group. Let

$$m_{ij} = 2(\log |C_{ij}| - \log b) \tag{6.2.6}$$

We can see from (6.2.5) and (6.2.4) that

$$m_{ij} \le \log \frac{a}{b} + 1 \tag{6.2.7}$$

By Claim 5.2.1 and (6.2.6), for each cluster C_{ij} of the second group there is a linear function

$$\varphi_{ij} : E^n \to E^{m_{ij}}$$

such that

$$I(\varphi_{ij}, C_{ij}) \le \frac{b^2}{|C_{ij}|} \tag{6.2.8}$$

The characteristics of φ_{ij} are:

$$\Pr(\varphi_{ij}) = n, \quad T(\varphi_{ij}) = n, \quad \text{PT}(\varphi_{ij}) = 2^n |C_{ij}| \cdot n \tag{6.2.9}$$

The function f_{ij} subdivides C_{ij} into $2^{m_{ij}}$ smaller subclusters C_{ijk}. Those subclusters C_{ijk} with the clusters C_{ij} of the first group constitute, by definition, the atoms of a partition B. We will prove that B meets the requirements of the Lemma.

We have the following upper bound for the size of B:

$$|B| \le \sum_i \left(2^{m_i} + \sum_j 2^{m_{ij}} \right) \tag{6.2.10}$$

Substitute into (6.2.10) equalities (6.2.1) and (6.2.6):

$$|B| \le \sum_i \left(\frac{|A_i|}{b} + \sum_j \frac{|C_{ij}|^2}{b^2} \right) \tag{6.2.11}$$

We obtain from (6.2.11), (6.2.2), (6.2.5) and an obvious equality

$$\sum |A_i| = |S|,$$

that

$$|B| \le \frac{2|S|}{b}(1 + o(1)) \tag{6.2.12}$$

The size of an atom of the first group is not greater than b by definition. The size of an atom of the second group is not greater than $b+1$ as it follows from Lemma 2.4.3 and (6.2.8).

Define two functions, H_1 and H_2, of two integer-valued variables i and j.

$H_1(i, j)$ is defined iff C_{ij} belongs to the first group of clusters. Then $H_1(i, j)$ is the quantity of those clusters $C_{i'j'}$, which, first, belong to the first group, and, second, for which $i' < i$.

$H_2(i, j)$ is defined iff C_{ij} is a member of the second group. Then $H_2(i, j)$ is the quantity of clusters $C_{i'j'k}$ such that either $i' < i$ or $i = i'$ and also $j' < j$.

Take a word $w \in S$. Let i be the number of the atom, which w belongs to:

$$i = \text{Mem }_A w.$$

The atom A_i was subdivided by φ_i into subclusters. Two cases may present themselves, $C_{i\varphi_i(w)}$ belongs to either first or second group.

In the first case

$$\text{Mem }_B(w) = H_1(i, \varphi_i(w)) + \varphi_i(w) \tag{6.2.13}$$

by the definition of H_1.

In the second case

$$\text{Mem }_B(w) = l + H_2(i, \varphi_i(w)) + \varphi_{i,\varphi_i(w)}(w) \tag{6.2.14}$$

by the definition of H_2.

The functions H_1 and H_2 are monotone with respect to the lexicographic order on the set of pairs (i, j). The jumps of H_1 are not greater than 2^{m_i}. The jump of H_2 at (i, j) is $2^{m_{ij}}$, which, by (6.2.6), is not greater than

$$2^{\log \frac{a}{b}} \le a.$$

The range of both those functions is $|B|$. Claim 2.3.4 and inequality (6.2.10) yield for the programlengths of H_1 and H_2:

$$\Pr(H_1) = O\left(\frac{|S|}{b}(\log \log |S| + \log a)\right) = \Pr(H_2) \tag{6.2.15}$$

The same Claim yields for the precomputing time and running time:

$$T(H_1) = O\left(\log(|B| \cdot a)\right) = T(H_2),$$

$$\text{PT}(H_1) = \text{O}(|B| \cdot a) = \text{PT}(H_2) \qquad (6.2.16)$$

In order to find Mem $_B$ we use either (6.2.13) or (6.2.14). We employ functions φ_i and φ_{ij}. The programlength of the map $A \to B$ transforming A to B equals

$$\text{Pr}(A \to B) = \text{Pr}(H_1) + \text{Pr}(H_2) + \sum_i \text{Pr}(\varphi_i) +$$

$$+ \sum_{i,j} \text{Pr}(\varphi_{i,j}) + \text{size [of the table of indices]} \qquad (6.2.17)$$

The quantity of functions φ_i is not greater than $|A|$. The quantity of functions φ_{ij} is not greater than $\sum 2^{m_i}$, which, by (6.2.12), does not exceed

$$\frac{2|S|}{b}(1 + \text{O}(1))$$

From (6.2.17), (6.2.3), (6.2.8) and (6.2.15) we get

$$\text{Pr}(A \to B) = \text{O}\left(\frac{|S|}{b}(n + l_2|S|)\right).$$

The running time is $\text{O}(n + \log|B| \cdot a)$, which equals $\text{O}(n + \log|S|)$ by (6.2.12). The precomputing time is

$$\text{O}\left(\frac{a^2}{b^2} \cdot n^3\right) + \sum_i \text{PT}(\varphi_i) + \sum_{i,j} \text{PT}(\varphi_{ij}) + \text{PT}(H_1) + \text{PT}(H_2) \qquad (6.2.18)$$

From (6.2.3), (6.2.9), (6.2.16) and (6.2.18) we obtain

$$\text{PT}(A \to B) = \text{O}(|S| \cdot 2^n \cdot n)$$

Q.E.D.

By way of example let us consider the dictionary $S = \{4, 16, 25, 36, 49, 64, 81, 100, 121, 144, 169, 196, 225, 216, 243\}$, $S \subseteq E^8$. The partition A consists of a sole atom S. We transform A into a partition B, $s(B) = 2$, $\gamma(B) = 2$, $|B| = 11$.

A word $w \in E^8$ is represented as $0w \in E^9$. By Lemma 6.2.1, we take a linear function φ_{421}. It breaks the dictionary S down into eight clusters $C_0, \dots, C_7$, which are shown on Table 6.2.1. Each of the clusters C_3, C_4, C_7 contains more than two words. Those clusters constitute the second group. The other clusters of the second group is given a linear function from E^8 to E^2, which is shown on Table 6.2.1. The clusters of the second group are broken down into smaller subclusters, which constitute a partition B. The sizes of those subclusters are not greater than 2. Functions H_1 and H_2 are displayed as well.

Table 6.2.1.

Atom splitting.

S	4,16,25,36,49,64,81,100,121,144,169,196,225, 216,243																
φ	φ_{421}																
C	121	216 81	-	100 169		196 243		4,16,64				225	49		25, 36, 144		
φ_i				φ_{1111}				φ_{1321}							φ_{1321}		
Δ	121	216 81	-	-	-	100 196	169 243	-	64	4	16	225	49	-	25 144	-	36
H_1	1	2	3	-	-	-	-	-	-	-	-	4	5	-	-	-	-
H_2	-	-	-	1	2	3	4	5	6	7	8	-	-	9	10	11	12

Lemma 6.2.2. (atom merging)

Let S be a set of n-length words, $A = \{A_1, \ldots, A_{|A|}\}$ be a partition of S, $s(A) = a$, b be a parameter. Then A is a subpartition of a partition B, $B = \{B_1, \ldots B_{|B|}\}$ such that

1. $ab \le |B_i| \le a(b+1)$, $i = 1, \ldots, |B|$, *i.e.*, $s(B) = a(b+1)$, $\gamma(B) \le a+1$

2. $PT(A \to B) = |A| \log a$

3. $\Pr(\text{Mem }_B) = \Pr(\text{Mem }_A) + |A|(1 + l_2 |A|)$.

4. $\Pr(\text{Prec }_B) = \frac{|S|}{ab}(\log ab + l_2 |S|)$

5. $T(\text{Mem }_B) = O(\log|S|) = T(\text{Prec }_B)$.

Proof.

Contiguous atoms of A are merged into atoms of a new partition B to make the sizes $|B_i|$ to meet the inequalities

$$ab \le |B_i| \le ab + a \qquad (6.2.19)$$

It is possible, since

$$|A_i| \le a, \quad i = 1, \ldots, |A|.$$

The function Mem $_B$ is a constant on an atom A_i, $i = 1, \ldots |A|$. It depends monotonically on the number i. Its jumps equal 1. We obtain by Claim 2.3.4 that the programlength of Mem $_B$ is

$$\Pr(\text{Mem }_B) = \Pr(\text{Mem }_A) + |A|(1 + \log\log|A|).$$

The programlength of Prec $_B$ is estimated the same way. This function depends monotonically on atom's number. Its jumps are not greater than $a(b+1)$, by (6.2.19). By Claim 2.3.4,

$$\Pr(\text{Prec }_B) = \frac{|S|}{ab}(\log ab + \log\log|S|).$$

The running times of Mem $_B$ and Prec $_B$ are $\mathrm{O}\left(\log |S|\right)$. Q.E.D.

Claim 6.2.1. (Galois multilevel partition)
Let S be a nontrivial source of n-length words,

$$\lim_{n \to \infty} \frac{\log |S|}{n} > 0.$$

There is a partition $A = \{A_1, \ldots A_{|A|}\}$ such that the maximal size of it atoms is

$$s(A) = \frac{1}{2} \log |S| + \mathrm{o}\left(\log |S|\right), \quad |A_i| = \frac{\log |S|}{2} + \mathrm{o}\left(\log |S|\right)$$

the quantity of different sizes is

$$\gamma(A) = \mathrm{o}\left(\log |S|\right).$$

The functions Mem $_A$ and Prec $_A$ have the programlengths equal to $\mathrm{o}\left(|S|\right)$, their running time is $\mathrm{O}\left(n\right)$, precomputing time is $\mathrm{O}\left(|S|^3 n\right)$.
Each word of S is one-to-one represented in an atom of A by a $7\log\log|S| + 1$-length binary word.

Proof.
If

$$n > 2\log |S| + C, \tag{6.2.20}$$

then take an injective function $g: E^n \to E^m$, where

$$m = 2\log |S| + C$$

(Corollary 5.3.4). The characteristics of g are:

$$\mathrm{Pr}\left(g\right) = \mathrm{O}\left(|S|^{2/3} \log^2 |S|\right), \quad T(g) = \mathrm{O}\left(n\right),$$

$$\mathrm{PT}\left(g\right) = \mathrm{O}\left(|S|^3 \cdot n\right) \tag{6.2.21}$$

Changing each word w for $f(w)$, we can consider the wordlength equal to m from the very beginning. If (6.2.20) is not met, the words of S remain unchanged. Anyhow, we can take the relation

$$n \leq 2\log |S| + C \tag{6.2.22}$$

as granted.
Let

$$a = |S|, \quad b = \log^2 |S|.$$

Apply Lemma 6.2.1 to the partition, whose sole atom is S. We will obtain a partition

$$B = \{B_1, \ldots B_{|B|}\}$$

such that

$$|B_i| \leq b+1, \quad i = 1, \ldots, |B|, \quad |B| \leq \frac{2|S|}{b}(1 + \mathrm{O}(1)),$$

$$T(\mathrm{Mem}\ _B) = \mathrm{O}\,(\log |S|) = T(\mathrm{Prec}\ _B),$$

$$\mathrm{Pr}\,(\mathrm{Mem}\ _B) = \mathrm{o}\,(|S|) = \mathrm{Pr}\,(\mathrm{Prec}\ _B),$$

$$\mathrm{PT}\,(\mathrm{Mem}\ _B) = \mathrm{PT}\,(\mathrm{Prec}\ _B) = \mathrm{O}\,(|S|^3 \log |S|) \qquad (6.2.23)$$

We have taken into account (6.2.22). Next let

$$a = \log^2 |S|, \quad b = \log |S|.$$

Apply Lemma 6.2.2 to the partition B. We will get a partition $C = \{C_1, \ldots, C_{|C|}\}$,

$$|C_i| = \log^3 |S|\,(1 + \mathrm{o}\,(1)) \qquad (6.2.24)$$

The characteristics of the partition C are of the same order of magnitude as the corresponding characteristics of B.

Apply Claim 5.3.1 to an atom C_i, $i = 1, \ldots, |C|$. We will get a polynomial f_b that is injective on C_i. The total time of selection of polynomials for all atoms is $\mathrm{o}\,(|S|^2)$. The length of the word b that determines the polynomial f_b, is

$$7 \log \log |S| + 1.$$

The total length of all such words is

$$7 \cdot |C| \cdot l_2 |S| = \mathrm{o}\,(|S|).$$

Change every word w of C_i for $f_B(w)$. The running time, using the table of indices, is $\mathrm{O}\,(\log |S|)$.

Use Lemma 6.2.1 again. This time we apply it to the partition C. The parameters are:

$$a = \log^3 |S|, \quad b = (\log \log |S|)^2.$$

We obtain a partition

$$D = \{D_1, \ldots, D_{|D|}\}, \quad |D_i| \leq (l_2 |S|)^2 + 1, \quad |D| = \mathrm{O}\left(\frac{|S|}{l_2(|S|^2)}\right).$$

The programlength of the map transforming C to D is $\mathrm{o}\,(|S|)$, its precomputing time is $\mathrm{o}\,(|S|^2)$, the running time is $\mathrm{O}\,(\log |S|)$.

To complete the proof, we use Lemma 6.2.2 with

$$a = (l_2 |S|)^2, \quad b = \frac{\log |S|}{2a}.$$

Applying it to D, we will obtain a partition

$$A = \{A_1, \ldots A_{|A|}\}, \quad |A_i| = \frac{\log |S|}{2}(1 + o(1)).$$

The programlength of the map transforming D to A is $o(|S|)$, the precomputing time is $o(|S|^2)$, the running time is $O(\log |S|)$, $\Pr(\text{Prec}) = o(|S|)$.

Summing all programlengths and precomputing times, used to develop the partition A we obtain the Claim. Q.E.D.

6.3 Partition-Numerator Interface

Universal numerators were developed by the brute force method of exhaustive search in Section 4.5. An opportunity to make such numerator in a less time consuming way comes with a good partition. It should be combined with a universal numerator of smaller dictionaries.

Suppose we are given a partition $A = \{A_1, \ldots, A_{|A|}\}$ of a set S, $S \subseteq E^n$, $n \geq 1$. Words of S are one-to-one represented in A by m-length binary words, $m \geq 1$. The maximal size of atoms of A is $s(A)$; atoms of $\gamma(A)$ different sizes are present in A. The loading factor is α, $0 < \alpha \leq 1$.

For an atom A_i take the numerator

$$U_{|A_i|}(E^m, \frac{1}{\alpha}|A_i|),$$

where

$$\frac{1}{\alpha}|A_i| = [0, 1, \ldots, \frac{1}{\alpha}|A_i| - 1].$$

This numerator contains a map

$$g : E^m \to [0, 1, \ldots, \frac{1}{\alpha}|A_i| - 1],$$

injective on A_i. Denote by $|\bar{U}|$ the union of all those numerators:

$$\bar{U} = \cup_{i=1}^{|A|} U_{|A_i|}(E^m, \frac{1}{\alpha}|A_i|) \qquad (6.3.1)$$

The set $\bar{U}$ may be called the numerator corresponding to A.

The cardinality $\left|U_{|A_i|}(E^m, \frac{1}{\alpha}|A_i|)\right|$ increases with $|A_i|$. Hence,

$$|\bar{U}| \leq \gamma(A) \left|U_{s(A)}(E^m, \frac{1}{\alpha}s(A))\right| \qquad (6.3.2)$$

Give each map in $\bar{U}$ a number from $[0, \ldots, |\bar{U}| - 1]$. The ranges of all maps may be considered equal to $[0, \ldots, \frac{1}{\alpha}s(A) - 1]$. Each map g has a table consisting of 2^m places.

There is a $\log \frac{1}{\alpha} s(A)$ - digit number at each place. So, the combined bitsize of all tables is

$$\Pr(\bar{U}) = 2^m \cdot \left| \bar{U} \right| \cdot \log \frac{1}{\alpha} s(A) \qquad (6.3.3)$$

The map g is injective on A_i. It is given a number within $\bar{U}$. Denote that number by Itr $_A(i)$. The interface function Itr $_A$ is determined by a table with $|A|$ places. There is Itr $_A(i)$ at the i-th place. So, the program complexity of Itr $_A$ is

$$\Pr(\text{Itr }_A) = |A| \log \left| \bar{U} \right| \qquad (6.3.4)$$

It takes one indexed table read to compute Itr, which is

$$T(\text{Itr}) = O(\log |A|) + \log |U|$$

bitoperations. To precompute the table we take every pair (A_i, f), where $A_i \in A$, $f \in \bar{U}$. Check whether f is injective on A_i. For that end we find all $f(w)$, $w \in A_i$, which takes $|A_i|$ table reads. Then we compare all words $f(w_i)$, $f(w_j)$, which takes $O(|A_i|^2)$ comparisons. So, the precomputing time of Itr is

$$\text{PT}(\text{Itr }_A) = |A| \cdot \left| \bar{U} \right| \cdot (s(A))^2 \cdot \log^2 s(A) \qquad (6.3.5)$$

Claim 6.3.1.

  Let A be a partition of a set $S \subseteq E^n$, $n \geq 1$, Mem $_A$ and Prec $_A$ be its membership and precedence functions, $\bar{U}$ be the corresponding numerator, Itr be the interface function, be a number, $0 < \alpha \leq 1$. Then there is a function

$$f : E^n \to [0, \dots, \frac{1}{\alpha} |S| - 1],$$

which is injective on S. Every one of the three characteristics of f (program complexity, running time, precomputing time) is the sum of the corresponding characteristics of Mem $_A$, Prec $_A$, $\bar{U}$, Itr $_A$ *and* Rep $_A$.

Proof.
Precompute the numerator $\bar{U}$ and the interface function Itr for a partition A. Define a map

$$f : E^n \to [0, \dots, \frac{1}{\alpha} |S|]$$

by the formula

$$f(w) = \frac{1}{\alpha} \left(\text{Prec }_A w + g_{\text{Itr }_A(\text{Mem }_A w)}(\text{Rep }_A w) \right). \qquad (6.3.6)$$

Here Mem $_A w$ is the number of the atom, which w belongs to; Rep $_A w$ is the representative of w in $A_{\text{Mem }_A w}$; $g_{\text{Itr }_A(\text{Mem }_A w)}$ is the function injective on $A_{\text{Mem }_A w}$; Prec $_A w$ is the quantity of words preceding w. When the functions Prec $_A$, Mem $_A$, Itr $_A$, Rep $_A$

and the set of functions $\bar{U}$ are known, we find $f(w)$ by (6.3.6). We make one addition and one division by α. The running time of those two operations is negligible. Q.E.D.

Claim 6.3.1 yields a method to develop a simple function injective on a dictionary. There is an analogous method to develop a simple partial boolean function.

Suppose f is a partial boolean function, S is its domain, $A = \{A_1, \ldots, A_{|A|}\}$ is a partition of S, Rep $_A w$ is the representative of w in A, $|\text{Rep } _A w| = m$. For an atom A_i take the piercing set $V_{|A_i|}(m)$ (Section 4.1). This set contains a boolean function g such that

$$g(\text{Rep } _A w) = f(w), \quad w \in A_i, \quad i = 1, \ldots, |A|. \tag{6.3.7}$$

Denote by $\bar{V}$ the union of all those piercing sets:

$$\bar{V} = \bigcup_{i=1}^{|A|} V_{|A_i|}(m) \tag{6.3.8}$$

$\bar{V}$ is called the piercing set corresponding to A. We get for its cardinality:

$$\left|\bar{V}\right| \le \gamma(A) \left|V_{s(A)}(m)\right| \tag{6.3.9}$$

Each map in $\bar{V}$ is given both a number and a table. The combined bitsize of all tables is

$$\text{Pr}\left(\bar{V}\right) = 2^m \cdot \left|\bar{V}\right| \tag{6.3.10}$$

The boolean function g meeting (6.3.7) on A_i is given a number within $\bar{V}$. We denote that number by Itr (i). The program complexity of the interface function Itr is

$$\text{Pr}\left(\text{Itr } _A\right) = |A| \cdot \log\left|\bar{V}\right| \tag{6.3.11}$$

Its precomputing time is

$$\text{PT}\left(\text{Itr } _A\right) = |A| \cdot \left|\bar{V}\right| \cdot s(A) \tag{6.3.12}$$

(Take every pair g, A_i, where $A_i \in A$, $g \in V$. Check whether or not g agrees with f on A_i).

Claim 6.3.2.

Let A be a partition of the domain S of a partial boolean function f, $S \subseteq E^n$, be the interface function, Mem $_A$ be the membership function of A, $\bar{V}$ be the corresponding piercing set, Itr $_A$ be the representation function. Then every one of the three characteristics of f (program complexity, running time, precomputing time) is the sum of the corresponding characteristics of Mem $_A$, $\bar{V}$ and Itr and Rep $_A$.

Proof.

A map f can be expressed as

$$f(w) = g_{\text{Itr}(\text{Mem } _A w)} \text{Rep } _A w \tag{6.3.13}$$

The statement of the Claim follows this formula.

Suppose that every subset $S \subseteq E^n$ of a given cardinality T, $|S| = T$, has got a partition A. Denote by $\Pr(\text{Mem})$, $T(\text{Mem})$, $\text{PT}(\text{Mem})$, $\Pr(\text{Prec})$, $T(\text{Prec})$, $\text{PT}(\text{Prec})$, $\Pr(\text{Rep})$, $T(\text{Rep})$, $\text{PT}(\text{Rep})$ the maximal value of the corresponding characteristic over all those partitions. We suppose that words of E^n are represented in those partitions by binary m-length words. We suppose that the same numerator $\bar{U}$ corresponds to all partitions. The maximal values of $\Pr(\text{Itr}_A)$, $\text{PT}(\text{Itr}_A)$ and $T(\text{Itr}_A)$ are denoted by $\Pr(\text{Itr})$, $\text{PT}(\text{Itr})$, $T(\text{Itr})$.

Claim 6.3.3.

Let $\bar{U}$ be a universal numerator corresponding to a partition of every dictionary S, $|S| = T$, $S \subseteq E^n$. Then there is a universal numerator U_1 of T-sized subsets of E^n, whose characteristics are:

$$\log_2 |U_1| = \Pr(\text{Mem}) + \Pr(\text{Prec}) + \Pr(\text{Itr}) + \Pr(\text{Rep});$$

$$\text{PT}(U_1) = |U_1| \cdot 2^n \cdot (T(\text{Prec}) + T(\text{Mem}) + T(\text{Itr}) + T(\text{Rep})) + \text{PT}(u).$$

The running time of any function from U_1 is

$$T(\text{Prec}) + T(\text{Mem}) + T(\text{Itr}) + T(\text{Rep}).$$

Proof.

Given a universal numerator u and four numbers $\Pr(\text{Mem})$, $\Pr(\text{Prec})$, $\Pr(\text{Itr})$ and $\Pr(\text{Rep})$, take one by one each quadruple of binary words a_1, a_2, a_3, a_4, whose lengths are just $\Pr(\text{Mem})$, etc., $\Pr(\text{Rep})$. For every partition A of a set S there is a quadruple such that a_1 is a program of the membership map Mem_A, etc., a_4 is a program of the representation map Rep_A. Load this word and a word w into a computer. After $T(\text{Mem})$ units of time we will get $\text{Mem}_A w$. Some words a_1 will yield $\emptyset$. The same is valid for Prec, Itr and Rep.

Formula (6.3.6) defines a map injective on a set S. Run over all quadruples (a_1, a_2, a_3, a_4) and all words w, $|w| = n$. Write down the results of applying (a_1, a_2, a_3, a_4), as programs, to w and calculating (6.3.6). We will get a set of tables of functions $E^n \to [0, \frac{1}{\alpha}T - 1]$. For any dictionary S there is a function injective on S in the set. It means that we obtain tables of function, which constitute a new numerator U_1 of all T-sized subsets S, $S \subseteq E^n$. There are $|U_1|$ quadruples, which is required by the Lemma. The running time of a quadruple on a word is not greater than

$$T(\text{Mem}) + T(\text{Prec}) + T(\text{Rep}) + T(\text{Itr}).$$

It yields the estimates of the Lemma for $\text{PT}(U_1)$.Q.E.D.

The following Claim is an analogue of Claim 6.3.3 for partial boolean functions.

Claim 6.3.4.
   Let $\bar{V}$ be a piercing set, which corresponds to a partition of every domain S, $|S| = T$, $S \subseteq E^n$. Then there is a piercing set $V_T^1(n)$, whose characteristics are

$$\log_2 \left|V_T^1(n)\right| = \mathrm{Pr}\,(\mathrm{Mem}\,) + \mathrm{Pr}\,(\mathrm{Prec}\,) + \mathrm{Pr}\,(\mathrm{Itr}\,) + \mathrm{Pr}\,(\mathrm{Rep}\,),$$

$$\mathrm{PT}\,(V_T^1(n)) = \left|V_T^1(n)\right| \cdot 2^n(T(\mathrm{Mem}\,) + T(\mathrm{Prec}\,) + T(\mathrm{Itr}\,) + T(\mathrm{Rep}\,)) + \mathrm{PT}\,(\bar{V}).$$

The running time of every boolean function from $V_T^1(n)$ is

$$T(\mathrm{Prec}\,) + T(\mathrm{Mem}\,) + T(\mathrm{Itr}\,) + T(\mathrm{Rep}\,).$$

The numbers $\mathrm{Pr}\,(\mathrm{Mem}\,)$ etc. are introduced before the Claim 6.3.3.

6.4 Low Redundant Numerators

Theorem 6.4.1.
   Let A and B be sets, T be natural, $\alpha = \frac{T}{|B|}$ be a constant, $\frac{T}{|A|} \to 0$. Then for every set S, $S \subseteq A$, $|S| = T$, $S \to B$ there is an injection f whose characteristics are:

1. $\mathrm{Pr}\,(f) = T\log_2 e\,\left(1 + \left(\frac{1}{\alpha} - 1\right)\ln\,(1 - \alpha)\right)(1 + \mathrm{o}\,(1)) + \mathrm{O}\,(\ln\ln|A|)$

2. $\mathrm{PT}\,(f) = \mathrm{O}\,(T^3\ln^{\,3}|A|)$

3. $T(f) = \mathrm{O}\,(\log|A|\log\log|A|)$

The program complexity of f is asymptotically minimal, the running time is minimal to within the factor $\log\log|A|$, the precomputing time is polynomial.
Proof.
Develop a partition A for a set S by Lemma 6.1.1. For that partition

$$s(A) = \mathrm{O}\,(l_2(n) \cdot l_3(n)), \quad |\mathrm{Rep}\,w| = m = \mathrm{O}\,(l_3(n)), \quad w \in S,$$

$$\gamma(A) = 1, \quad \mathrm{Pr}\,(\mathrm{Mem}\,_A) = \mathrm{Pr}\,(\mathrm{Prec}\,_A) = \log n(1 + \mathrm{o}\,(1)) + \mathrm{o}\,(T),$$

$$n = \log|A|, \quad T = |S|. \tag{6.4.1}$$

Develop a universal numerator $\bar{U}$ by Corollary 4.5.1. We obtain from (6.3.2):

$$\ln\left|\bar{U}\right| = s(A)\left(1 + \left(\frac{1}{\alpha} - 1\right)\ln\,(1 - \alpha)\right)(1 + \mathrm{o}\,(1))$$

$$\mathrm{Pr}\,(\bar{U}) = \mathrm{o}\,(\log n), \quad \mathrm{PT}\,(\bar{U}) = \mathrm{O}\,(n) \tag{6.4.2}$$

From (6.3.4) we get

$$\Pr\left(\text{Itr}\right) = \frac{T}{s(A)} \cdot \log\left|\bar{U}\right| = T\left(1 + \left(\frac{1}{\alpha} - 1\right)\ln\left(1 - \alpha\right)\right)\left(1 + \text{o}\left(1\right)\right) \qquad (6.4.3)$$

The precomputing time is given by (6.3.5):

$$\text{PT}\left(\text{Itr}\right) = \text{o}\left(|S|^2\right) \qquad (6.4.4)$$

Combine the partition A with the numerator $\bar{U}$ through Claim 6.3.1. It yields the Theorem. Q.E.D.

Take the partition A of table 6.1.1 and the universal numerator $U_2(E^2, E^1) = \{f_0, f_1\}$ (Section 4.5). The function f_1 is injective on A_0 and A_3, f_0 – on A_0, A_1, A_2. An injection f is given by (6.3.6):

$$f(0) = 1, \quad f(1) = 0, \quad f(2) = 2, \quad f(6) = 4, \quad f(7) = 5, \quad f(8) = 3.$$

Another example. Let $A = E^4$, $B = \{0, 1, 2\}$, $S = \{0, 1, 2\}$ (decimal). Take the function f_2 from the set $G(4, 2)$ (Table 5.2.1). It transforms words of S into their representatives: $00, 10, 11$. The function F_1 from the set $U_3(A, B)$ (section 4.5) maps those representatives into $1, 2, 0$. The composition $F_1 \circ f_2$ is injective on S.

Claim 6.4.1.
Let A and B be sets, T be an integer, $\alpha = \frac{T}{|B|}$ be a constant, $\frac{T}{|A|} \to 0$. There is a universal numerator $U_T(A, B)$, whose characteristics are

$$\log |U_T(A, B)| = T \log_2 \text{e} \left(1 + \left(\frac{1}{\alpha} - 1\right)\ln\left(1 - \alpha\right)\right)\left(1 + \text{o}\left(1\right)\right) + \text{O}\left(\ln\ln|A|\right)$$

$$\text{PT}\left(U_T(A, B)\right) = \text{O}\left(|U_T(A, B)| \cdot 2^n \cdot n \cdot \ln n\right).$$

The running time of functions from $U_T(A, B)$ is $\text{O}\left(n \log n\right)$, where $n = \log|A|$.
Proof.
Lemma 6.1.1 provides every dictionary S with a partition A. The maximal over all A values of corresponding characteristics are:

$$\Pr\left(\text{Mem}\right) = \Pr\left(\text{Prec}\right) = \log n(1 + \text{o}\left(1\right)) + \text{o}\left(T\right),$$

$$\text{PT}\left(\text{Mem}\right) = \text{PT}\left(\text{Prec}\right) = \text{O}\left(T^3 n^2\right)$$

$$s(A) = \text{O}\left(l_2(n) \cdot l_3(n)\right) \qquad (6.4.5)$$

The words of S are represented in A by $m = \text{O}\left(l_3(n)\right)$-length words.

The running time of functions Mem and Prec is $\text{O}\left(n \log n\right)$.

Develop a universal numerator U by Corollary 4.5.1 for every partition A. Combining U with A by Claim 6.3.3 gives a new universal numerator $U_T(A, B)$, whose characteristics are as claimed.

The running time of functions in Theorem 6.4.2 is lowered by a factor $\log \log |A|$ as compared with Theorem 6.4.1. Theorem 6.4.2 is applicable to nontrivial sources $\left(\lim \frac{\ln T}{\ln |A|} > 0 \right)$ only.

Theorem 6.4.2.

Let A and B be sets, T be natural, $\alpha = \frac{T}{|B|}$ be a constant. Consider nontrivial sources $\left(\lim \frac{\ln T}{\ln |A|} > 0 \right)$. Then for every set S, $S \subseteq A$, $|S| = T$, there is an injective function $f : S \to B$, whose characteristics are:

1. $\Pr(f) = T \log_2 e \left(1 + \left(\frac{1}{\alpha} - 1 \right) \ln (1 - \alpha) \right) (1 + o(1))$

2. $\mathrm{PT}(f) = O(T^3 n)$

3. $T(f) = O(\log |A|)$

Both the program complexity and the running time of f are minimal.

Proof.
The proofs of Theorem 6.4.1 and 6.4.2 are analogous. There are two points of difference. First, the Galois multilevel partitions of Claim 6.2.1 are taken in Theorem 6.4.2 instead of lexicographic-polynomial partitions in theorem 6.4.1. Second, universal numerators of Claim 6.4.1 are taken in theorem 6.4.2 instead of Corollary 4.5.1 in Theorem 6.4.1. As a result we obtain the fastest injective functions. Their running time is $O(\log |A|)$. Q.E.D.

6.5 Highly Redundant Numerators

The numerator of dictionaries S, $|S| = T$ with a constant loading factor is called low redundant, $T \to \infty$. The numerator $f : A \to B$ with a constant redundancy ρ,

$$\rho = \frac{\log |B| - \log T}{\log T},$$

is called highly redundant. The more redundancy, the less complexity. Identity mapping is the simplest, but it is overredundant. We will find redundancy – complexity trade-off for nontrivial sources $\left(\lim \frac{\log T}{\log |A|} > 0 \right)$.

Theorem 6.5.1.

For any T, A, B, ρ, where $\lim \frac{\log T}{\log |A|} > 0$, $\frac{\log |B| - \log T}{\log T} = \rho$, there is a universal numerator $U_T(A, B)$, whose cardinality is minimal:

$$\log |U_T(A, B)| = \begin{cases} T^{1 - \rho + o(1)}, & \rho < 1 \\ O(\log |A|), & \rho > 1 \end{cases}$$

The time to compute $f(x)$, $f \in U_T(A, B)$, $x \in A$, is $O(\log T \cdot \log \log |A|)$. The time to precompute for $S \subseteq A$ an injective function $f \in U_T(A, B)$ is $O(T(\log T)^C)$, $C = $ const .

Proof.
The minimality of the universal numerator follows Corollary 4.5.2. Next we construct such a numerator avoiding exhaustive search.

First, let $\rho < 1$. Choose ρ_1, $0 < \rho_1 < \rho < 1$. Employ Corollary 5.3.2 to build a $T^{-\rho_1+o(1)}$ -hash set $U_T(A, B, T^{-\rho_1+o(1)})$. The building time is $O(T(\log T)^C)$.

For a dictionary $S \subseteq A$, $|S| = T$, there is a map $f' \in U_T(A, B)$, $I(f', S) \leq T^{-\rho_1+o(1)}$. The map f' is polynomial. Its program length is

$$\Pr(f') = O(n) = o(T^{1-\rho}) \tag{6.5.1}$$

by the condition of nontriviality $\lim \frac{\log T}{n} > 0$.

The dictionary S consists of two subdictionaries, S_1 and S_2. The words of S_1 are colliding, whereas the words of S_2 are not. By Lemma 2.4.1,

$$|S_1| \leq T \cdot T^{-\rho_1+o(1)} = T^{1-\rho_1+o(1)} \tag{6.5.2}$$

Use Theorem 5.7.1. It takes time

$$O(|S_1|(\log|A|)^2) = O(T(\log T)^C), \quad C = \text{const},$$

to precompute for S_1 a lexicographic map f_1, which gives a number to every $x \in S_1$. If $x \notin S_1$, then $f_1(x) = \emptyset$. The time to compute f_1 is as claimed.

We define f to be equal either the concatenation $1f_1$, if $f_1 \neq \emptyset$, or $0f'$, if $f_1 = \emptyset$, $x \in S_1$. The program length of f is

$$|S_1| \cdot \log|S_1| + \Pr(f') = O(T^{1-\rho}).$$

It means that the cardinality of the of all functions f meets the Theorem. The redundancy of f is $\rho_1 + o(1) \leq \rho$.

If $\rho > 1$, we define

$$1 < \rho_1 < \rho, \quad \mu = \lceil(1+\rho_1)\log T\rceil$$

and take the polynomial map f_b by (5.3.1). We get

$$I(f_b, S) \leq T^{-\rho_1+1} \leq \frac{1}{T}.$$

By Lemma 2.4.2, f_b is an injection. The program complexity of f_b is $O(n)$, i.e., the cardinality of $U_T(A, B)$ is $O(n) = O(\log|A|)$.

Corollary 6.5.1. (Entropy Doubling Effect).

The complexity to enumerate almost any nontrivial source of n-length words S $\left(h(S) = \frac{\log|S|}{n} > 0\right)$ by l-length words is exponential in n, if

$$l > 2(h(S) + \varepsilon).$$

It is linear, if

$$l < 2(h(S) - \varepsilon).$$

Proof.
It follows the Theorem immediately. The quantity of sources with a simpler enumeration is negligible, as it can be seen from Claim 4.5.1.

6.6 Channels with Arbitrary Additive Noise

Such a channel is specified by a set S of a binary n-length words (noises), $n > 1$.

The word $0 = (0, \ldots, 0)$ belongs to S. If an input to the channel is $w \in E^n$, then the output is $w + x$, $x \in E^n$. A set $K \in E^n$ is a code correcting S, if for any $y_1, y_2 \in K$ and any $x_1, x_2 \in S$

$$y_1 + x_1 \neq y_2 + x_2.$$

It means that if only words of K can enter the channel, then they can be identified by the output, notwithstanding the noise. The code K corrects noise (errors) from S.

It is a rather general scheme. It includes many interesting cases. For instance, if S is a ball with respect to the Hamming distance, then we get the binary symmetric channel. The code K corrects r symmetric errors in that case, r being the radius of S.

If K corrects S, then the sets $x + S$, where $x \in K$, are not interesecting. The cardinality of each of them is $|S|$. It yields:

$$\left| \bigcup_{x \in K} (x + S) \right| = |K| \cdot |S| \leq 2^n \qquad (6.6.1)$$

Taking logarithms:

$$h(K) + h(S) \leq 1. \qquad (6.6.2)$$

Here $h(S)$ is the Hartley per letter entropy of K, $h(S)$ is one of S. Inequality (6.6.2) is a generalization of the Hamming sphere packing inequality.

A set K is called a group code, if it is the set of zeroes of a linear map $f : E^n \to E^m$, $m > 0$. A group code corrects S iff the corresponding map f is injective on S. The image $f(x)$ of a word $x \in E^n$ is called its syndrome, $m - \log|S|$ is called the number of correcting digits.

Suppose a word $x \in K$ is an input, a word $x + y$, where $y \in S$, is an output. The syndrome of the output is

$$f(x + y) = f(y).$$

We find the noise y by the syndrome, because f is injective. We find x via y.

Let $A = E^n$, S be a nontrivial set $\left(\lim \frac{\log T}{n}\right) > 0$, $|B| \geq T^2 \cdot n$. For every S there is a linear map

$$f : E^n \to E^m, \quad m = \log |B|,$$

that is injective on S (Claim 5.3.1). For the number m we get

$$m \leq (2 + \varepsilon) \log |S|.$$

The set of zeroes f is a group code K, for which

$$h(K) \geq 1 - (2 + \varepsilon)h(S). \tag{6.6.3}$$

So, for any S there is a group code K, meeting (6.6.3), which is called the generalized Varshamov-Gilbert inequality. It was proved in Desa (1965).

Is there a code outperforming Varshamov-Gilbert's? It is a main information-theoretical problem, which is rather hard to solve in any particular case. For a typical noise, the Varshamov-Gilbert code is the best.

Claim 6.6.1.
For almost every noise S, $0 < h(S) < \frac{1}{2}$, and $\varepsilon > 0$ any group code with $m - \log |S|$ correcting digits does not correct S, if

$$m < (2 - \varepsilon) \log |S|.$$

Proof.
Let f be a linear map

$$f : E^n \to E^m, \quad m < (2 - \varepsilon) \log |S|.$$

We get for the redundancy $\rho(f, S)$ of f on S :

$$\rho(f, S) < 1 - \varepsilon. \tag{6.6.4}$$

As it follows from the Claim 4.5.1 in its asymptotical form (Corollary 4.5.2), the complexity $\Pr(f)$ of any injection $f : E^n \to E^m$ meets the inequality

$$\Pr(f) \geq |S|^{1-\rho} \geq |S|^{\varepsilon} \tag{6.6.5}$$

for almost any dictionary S. On the other hand, f is a linear map. It is specified by a $n \times m$ binary matrix, which is its computing program. Hence,

$$\Pr(f) \leq n \cdot m \leq n^2. \tag{6.6.6}$$

From (6.6.5), (6.6.4) and the nontriviality condition $(h(S) > 0)$, we obtain

$$n^2 \geq \Pr(f) \geq 2^{C \cdot n}$$

– a contradiction. Q.E.D.

Claim 6.2.2. (Generalized Shannon Theorem)

For any additive noise S and $\varepsilon > 0$ there is a group code K that corrects almost any noise from S. The number of correcting digits does not exceed $\varepsilon \log |S|$.

Proof.

Choose $\rho_1 < \varepsilon$. In the proof of Theorem 6.5.1 we developed a map

$$f' : E^n \to T^{1+\rho_1}, \quad I(f', S) \leq T^{-\rho_1 + o(1)}, \quad |S| = T.$$

Let K be the set of zeroes of f'. The number of correcting digits of K is

$$\log T(1 + \rho_1) - \log T \leq \varepsilon T,$$

as claimed. The number of colliding words $|S_1|$ does not exceed

$$T^{1 - \rho_1 + o(1)} = O(T),$$

i.e., almost any word of S is not colliding. Colliding words are just those noises not corrected by K. Q.E.D.

6.7 Partial Boolean Functions, Piercing and Independent Sets, Threshold Formulas

Claim 6.4.1 gives a set of tables for the functions of a universal numerator $U_T(A, B)$. The precomputing time cannot be less than

$$V = |U_T(A, B)| \times 2^n.$$

The precomputing time by Claim 6.4.1 is just slightly greater, equalling $V \cdot n \ln n(1 + o(1))$.

Each universal enumerator corresponds to a threshold formula (Section 4.5). So, to make a formula for $Th_T(x_1, \ldots x_N)$, $N = 2^n$, with

$$N e^{T \cdot (1 + o(1))}$$

occurrences of variables, we spend nearly minimal time $O(V \cdot \log V \cdot \log \log V)$.

Table 6.7.1 provides an example of such a formula. It is a continuation of the second example illustrating Theorem 6.4.1. We had $A = E^4$, $N = 16$, $T = 3$. The set $G(4, 2)$ is given by Table 5.2.1, the set $U_3(E^2, [0, 1, 2])$ consists of two functions. A function from $G(4, 2)$ transforms 4-bits words into their 2-bit representatives. A function from $U_3(E^2, [0, 1, 2])$ transforms a representative into a number $0, 1$ or 2. Taking all compositions $F \circ f$, where $F \in U_3(E^2, [0, 1, 2])$, $f \in G(4, 2)$, we will get all 8 functions from $U_3(E^4, [0, 1, 2])$. A program (binary number) of a function from that set consists of three bits. The first bit is the number of a function from $U_3(E^2, [0, 1, 2])$. The second and third bits are the number of a function from $G(4, 2)$. Say, the function number $5 = 101$ is the composition of F_1 and function number one from $G(4, 2)$. Its value on, say, 13 is $F_1(f_1(13)) = F_1(2) = 2$. We give the sets of values of functions from $U_3(E^4, [0, 1, 2])$ on all vectors from E^4.

Table 6.7.1.

functions	vectors															
	0	1	2	3	4	5	6	7	8	9	10	11	12	13	14	15
000	0	0	0	0	1	1	1	1	0	0	0	0	2	2	2	2
010	1	1	1	1	0	0	0	0	2	2	2	2	0	0	0	0
001	0	1	0	2	1	0	2	0	0	2	0	1	2	0	1	2
101	1	0	2	0	0	1	0	2	2	0	1	0	0	2	0	1
010	0	0	2	1	1	2	0	0	0	0	1	2	2	1	0	0
110	1	2	0	0	0	0	2	1	2	1	0	0	0	0	1	2
011	0	2	1	0	1	0	0	2	0	1	2	0	2	0	0	1
101	1	0	0	2	0	2	1	0	2	0	0	1	0	1	2	0

Functions from $U_3(E^4, [0, 1, 2])$
The conjunction
$$(x_0 \vee x_1 \vee x_2 \vee x_3 \vee x_8 \vee x_9 \vee x_{10} \vee x_{11}) \cdot (x_4 \vee x_5 \vee x_6 \vee x_7) \cdot (x_{12} \vee x_{13} \vee x_{14} \vee x_{15})$$
corresponds to function number 000.

Claim 6.3.4 gives a method to make a piercing set. Use Lemma 6.1.1 to partition each set $S \subseteq E^n$. Owing to Claim 6.3.4 we obtain

Claim 6.7.1.

There is a piercing set $V_T(n)$, whose cardinality meets the equality

$$|V_T(n)| = T(1 + o(1)) + \log n.$$

The precomputing time of $V_T(n)$ is

$$|V_T(n)| \cdot 2^n \cdot n \log n.$$

There is a relation between piercing sets and families of independent sets (Section 4.7). This relation yields

Claim 6.7.2.

A family F of T-independent subsets of an M-sized set, where $|F| = 2^{C_T \cdot M}$, $C_T = 2^{-C \cdot T}$, may be constructed in time $V \log V \log \log V$, where $V = |F| \cdot M$ is a lower bound for that time.

A method to compute a partial boolean function f is prescribed by Claim 6.3.2. We shall combine a partition A of the domain S with a piercing set. There are two options. The first is to take a lexicographic-polynomial partition and an exhaustive search piercing set. Then we get a minimal length program and $O(n \log n)$ running time. The second is to get a multilevel Galois partition and a piercing set of Claim 6.7.1. Then we get a minimal length program again. However, the running time will be $O(n)$. The condition $\lim_{n \to \infty} \frac{\ln |S|}{n} > 0$ should be met.

Theorem 6.7.1.

For every partial boolean function of n variables f, $|\mathrm{dom}\, f| = S$, there is a computing program, whose length is $T(1 + o(1)) + \log n$. The running time is $O(n \log n)$. It is diminished to $O(n)$, if $\lim_{n \to \infty} \frac{\log |S|}{n} > 0$. The precomputing time is $O(T^3 \cdot n)$.

Table 6.7.2 shows a partial boolean function f of four variables. Four-dimensional vectors are represented by integers from 0 to 15.

Table 6.7.2.
A partially defined boolean function

0	1	2	3	4	5	6	7	8	9	10	11	12	13	14	15
			1	0					1		1			1	0

The domain of f is $S = \{3, 4, 9, 11, 14, 15\}$. The vector 10 (binary notation is 1010) divides S into atoms A_1 and A_2,

$$A_1 = \{3, 4, 9\}, \quad A_2 = \{11, 14, 15\}.$$

For each atom there is a map in $G(4, 2)$ (Table 5.2.1), which is injective. It is f_0 for A_1 and f_2 for A_2. Those maps transform the vectors to their representatives:

$$3 \to 0, \quad 4 \to 1, \quad 9 \to 2, \quad 11 \to 3, \quad 14 \to 0, \quad 15 \to 2.$$

We get two partial functions of two variables. They are $01 \to 0$, $10 \to 1$ on A_1, $11 \to 1$, $00 \to 1$, $10 \to 0$ on A_2. Go to Table 4.7.1. As we can see, function number 010 agrees with f on A_1, 011 – on A_2. To calculate $f(w)$, compare w with 1010 and go to A_1 or A_2. Transform w by f_0 or f_1. Calculate function number 010 on A_1, number 011 – on A_2. We get $f(w)$.

6.8 Short Tables for Rapid Computation of Lipschitz Functions

The Lipschitz functional space $F(L, \Delta, C)$ was introduced in Section 1.2. It is the set of functions meeting the Lipschitz condition:

$$|f(x_1) - f(x_2)| \leq L|x_1 - x_2|, \quad x_1, x_2 \in \Delta.$$

The value assumed by a function $f \in F(L, \Delta, C)$ at the left end of Δ is within the segment $[-C, C]$. The ε-entropy of $F(L, \Delta, C)$ equals asymptotically $\frac{|\Delta|L}{\varepsilon}$.

An optimal ε-net was defined in Section 1.2. It consists of piecewise linear functions f_ε. To calculate a function f_ε we need, first, its initial value $f_\varepsilon(0)$, and, second, a code of the function $f_\varepsilon - f_\varepsilon(0)$. There are two very simple tables for such a function.

The first one is the $\frac{|\Delta|L}{\varepsilon}$-length binary codeword, given to it in Theorem 1.2.2. To reconstruct $f_\varepsilon - f_\varepsilon(0)$ by that word we should go from one $\frac{\varepsilon}{L}$-length subsegment to the next one, adding each time $\pm\varepsilon$. It takes $O\left(\log\frac{1}{\varepsilon} \cdot \frac{1}{\varepsilon}\right)$ bitoperations.

The second one is just the set of values assumed by $f_\varepsilon - f_\varepsilon(0)$ at points $\frac{k\varepsilon}{L}$, $k = 1, \ldots, \frac{|\Delta|L}{\varepsilon}$. To compute $f_\varepsilon\left(\frac{k\varepsilon}{L} - f_\varepsilon(0)\right)$, we should take it back from the k-th location of the table. In other words, the computation is simply one read in a table with $\frac{|\Delta|L}{\varepsilon}$ locations. It takes $O\left(\log\frac{1}{\varepsilon}\right)$ bitoperations. The size of the table is $O\left(\frac{1}{\varepsilon}\log\frac{1}{\varepsilon}\right)$ bits.

So, the first table is the shortest one asymptotically, although the computation time is far from the minimum. The other table is just the other way round. It is rather lengthy, but easy to operate with. The computation time is minimal.

Next we are going to construct a new table, which combines good qualities of both. It is asymptotically as short as the first one. The computation by the new table is as quick as by the second one.

The construction is as follows.

Given a segment Δ, constants L and C, subdivide Δ into a_1-length first level segments. Subdivide each first level subsegment into a_2-length second level subsegments. The numbers a_1 and a_2 will be fixed later.

A universal set of tables was developed in Section 1.2. For any function g that belongs to the ε-net F'' of $F(\Delta, L, 0)$ there is a table in the set through which one can compute g. The running time is $O\left(\log\frac{1}{\varepsilon}\right)$, the size of the set is given by (1.2.9).

Let f be a function from $F(\Delta, L, C)$, $\varepsilon > 0$, f_ε be the piecewise linear function, which approximates f with ε-accuracy. The new table for f_ε is composed of the first and second level initial tables, the interface table and a universal set of tables for Lipschitz functions restricted to the a_2-length subsegment. We will describe each of those tables.

Let $D = [D_l, D_r]$ be a first level subsegment, D_l and D_r be its left and right endpoints, $d = [d_l, d_r]$ be a second level subsegment,

$$D_r - D_l = a_1, \quad d_r - d_l = a_2.$$

The first level initial table provides any first level subsegment D with

$$f_\varepsilon(D_l) - f_\varepsilon(0).$$

The second level initial table provides any second level subsegment d with

$$f_\varepsilon(d_l) - f_\varepsilon(D_l),$$

where D is the first level subsegment, which d belongs to:

$$d \subseteq D, \quad D_l \leq d_l \leq d_r \leq D_r.$$

The sizes of those tables are, correspondingly,

$$\frac{|\Delta|}{a_1} \log \frac{2L \, |\Delta|}{\varepsilon} \tag{6.8.1}$$

bits and

$$\frac{|\Delta|}{a_2} \log \frac{2La_1}{\varepsilon} \tag{6.8.2}$$

bits. Restrict the function f_ε to a second level subsegment d. The function $f_\varepsilon - f_\varepsilon(d_l)$ coincides with a function g from the universal set of Lipschitz functions restricted to an a_2-length subsegments. The table of g is given a $\frac{\Delta L}{\varepsilon}$-length binary number. The interface is the table, which provides any second level subsegment d with this very number. There are $\frac{|\Delta|}{a_2}$ second level subsegments. Hence, the size of the interface program is

$$\frac{|\Delta|}{a_2} \cdot \frac{La_2}{\varepsilon} = \frac{L \, |\Delta|}{\varepsilon} \tag{6.8.3}$$

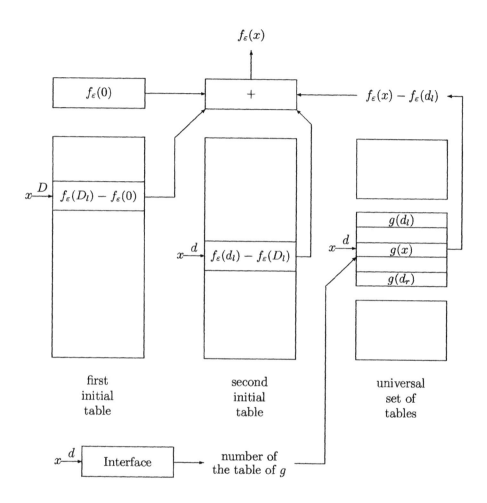

Fig. 6.8.1. *Computation of a Lipschitz function f. f_ε is an ε-approximation of f. g equals $f_\varepsilon - f_\varepsilon(0)$ on a subsegment d.*

bits. Fig. 6.8.1 shows the process of computation. Given $x \in \Delta$, we know the first level subsegment D and the second level subsegment d which x belongs to. Enter the first and second level initial tables. They return

$$f_\varepsilon(D_l) - f_\varepsilon(0) \quad \text{and} \quad f_\varepsilon(d_l) - f_\varepsilon(D_l).$$

Enter the interface table. It returns the number given in the universal set to the function g, which coincides with $f_\varepsilon - f_\varepsilon(d_l)$ on d. Find in the universal set necessary table and enter it. It returns $f_\varepsilon(x) - f_\varepsilon(d_l)$. Sum three numbers found and $f_\varepsilon(0)$. The number $f_\varepsilon(x)$ sought for is obtained. The running time is $O\left(\log \frac{1}{\varepsilon}\right)$.

To combined size of the initial tables, interface table, universal set and the number $f_\varepsilon(0)$ equals, by (6.8.1)-(6.8.3) and (1.2.9), to

$$\frac{|\Delta|}{a_1} \log \frac{2L\,|\Delta|}{\varepsilon} + \frac{|\Delta|}{2} \log \frac{2La_1}{\varepsilon} + \frac{L\,|\Delta|}{\varepsilon} + 2^{\frac{La_2}{\varepsilon}} \cdot \frac{La_2}{\varepsilon} \log \frac{2La_2}{\varepsilon} + o\left(\frac{1}{\varepsilon}\right) \qquad (6.8.4)$$

Here $|\Delta|$ and L are constants, ε goes to zero.

Let

$$\frac{La_2}{\varepsilon} = \log \frac{L\,|\Delta|}{\varepsilon} - 2l_2\left(\frac{L\,|\Delta|}{\varepsilon}\right), \quad l_2 = \log\log, \quad a_1 = O\left(\varepsilon \log^2 \frac{1}{\varepsilon}\right).$$

Then the third addend in (6.8.4) equals

$$\frac{L\,|\Delta|}{\varepsilon} = H_\varepsilon(F(\Delta, L, C)).$$

Any other addend is $o\left(H_\varepsilon(F(\Delta, L, C))\right)$, i.e., the size of the table is H_ε asymptotically. The length of $f_\varepsilon(0)$ is $O\left(\log \frac{1}{\varepsilon}\right)$. The above procedure gives an exact value of f_ε at the points $\frac{k\varepsilon}{L}$, $k = 0, 1, \ldots, \frac{\Delta L}{\varepsilon}$. Given $x \in \Delta$, find a number k, such that

$$\frac{k\varepsilon}{L} \leq x \leq \frac{(k+1)\varepsilon}{L}.$$

Suppose we know

$$\log \frac{|\Delta|\,L}{\varepsilon} + l_2\left(\frac{|\Delta|\,L}{\varepsilon}\right) = O\left(\frac{1}{\varepsilon}\right) \qquad (6.8.5)$$

bits of x. Then we know x with the accuracy

$$x - \frac{k\varepsilon}{L} = O\left(\frac{\varepsilon}{\log \frac{1}{\varepsilon}}\right) = o\left(\varepsilon\right) \qquad (6.8.6)$$

The function f_ε being piecewise linear, we have

$$\left| f_\varepsilon(x) - f_\varepsilon\left(\frac{k\varepsilon}{L}\right) \right| = L\left(x - \frac{k\varepsilon}{L}\right) \qquad (6.8.7)$$

From (6.8.6) and (6.8.7) we get, that we can calculate $f_\varepsilon(x)$ with the accuracy $o(\varepsilon)$. The given function is approximated by f_ε with ε-accuracy everywhere (Theorem 1.2.2). Now, we come to the following conclusion

Theorem 6.8.1.
Let f be a function with a Lipschitz constant L on a segment Δ, $\varepsilon \to 0$. There is a $\frac{\Delta L}{\varepsilon}(1 + o(1))$ sized table, which yields the values of f with $\varepsilon(1 + o(1))$-accuracy. The running time is $O\left(\log \frac{1}{\varepsilon}\right)$. Both the size of the table and the time are asymptotically optimal over the class of Lipschitz functions.

NOTES

A function injective on a dictionary is called a perfect hashing function.

Practical algorithms of perfect hashing are developed in Chang and Lee (1986), Cormac (1981), Fox et al. (1992) and in many other papers. Krichevskii (1976) set the problem theoretically. There is an algorithm of perfect hashing with program complexity $\log_2 e$ bits per dictionary word in that paper. This algorithm has two drawbacks. First, the running time is $O(\log^2 |A|)$, A being the universum. It is not optimal. Second, the algorithm cannot be applied to small dictionaries. In Fredman et al. (1984) the running time was reduced to $O(\log |A| \cdot \log \log |S|)$, while the program complexity did not reach the optimum. In Schmidt and the Siegel (1990) the program complexity went down to $O(|S| + \log \log |A|)$, which is optimal to within a constant factor. Still the exact optimum was not reached in that paper.

We present in this book a solution of the perfect hashing problem that can be considered final. The only room for furtner improvement: first, to lower the pre-computing time, second, to lower the running time from $O(\log |A| \log \log |A|)$ to $O(\log |A|)$ for small dictionaries.

Interesting variants of perfect hashing, dynamic and nonoblivious retrieval, are discussed in Dietzfelbinger et al. () and Fiat et al. (1992).

The idea of numerator-partition interface goes back to Shannon (1949) with con-tact trees playing the part of partitions and universal contact networks playing the part of numerators.

The channels with arbitrary additive noise were studied first by M. Desa (1965) and V. Goppa (1967).

Partial boolean functions are the subject of the papers Nechiporuk (1966), Pip-penger (1977), Sholomov (1969), Andreev (1989). A partial boolean function was given a minimal length program in those papers. However, the precomputing time was not polynomial.

We give a function a program of minimal length and minimal running time. The precomputing time is polynomial. The relation between the cardinality of the domain and the number of variables is arbitrary.

The results on T-independent sets are improvements of theorems from Kleitman and Spencer (1973), Alon (1986), Poljak et al. (1983).

Optimal tables for Lipschitz functions are constructed by V. Potapov and the author.

Appendix 1

Universal Numerators and Hash-sets

Universal numerator $U_T(A, B)$ for any T-sized subset S of A contains a map $f : A \to B$, which is injective on A. Universal a-hash set $U_T(A, B, a)$ for any T-sized subset S of contains a map $f : A \to B$, whose index on S is not greater than a. The logarithm of the size of a numerator is program complexity. The time is given to within a multiplicative constant.

Name	Size of the numerator	Precomputing time	Running time	Index guaranteed
Unordered	$|A|^T$	0	$T\log|A|$	0
Lexicographic Sect. 5.7	$|A|^{T(1+o(1))}$	$T(\log|A|)^2$	$\log T \log\log|A|$	0
Two-step Digital Sect. 5.5	$(\log|A|)^{CT}$	$T(\log|A|)^3$	$\log T\log|A|$	0
Enumerative	T^{CT}	$T(\log|A|)^2$	$\log T\log|A|$	0
Galois Linear Sect. 5.2	$|A|$	$|B|\,|A|\log|A|\log|B|$	$\log|A|\log|B|$	$T/|B|$
Galois Polynomial Sect. 5.3	$|B|$	$T\,|B|\log|A|\log|B|$	$\log|A|\log|B|$	$\frac{T\log|A|}{|B|\log|B|}$
Obtained by Exhaustive Search Sect. 4.5, $\frac{\ln\ln|A|}{T} \to 0$	$e^{\,T(1+o(1))}$	$T^{|A|(1+o(1))}$	$\log|A|$	0
Optimal	$e^{\,T(1+o(1))} \cdot \ln|A|$	$T^3(\ln|A|)^3$	$\log|A|\log\log|A|$	0
Optimal $T \geq \log^2|A| \log\log|A|$	$e^{\,T(1+o(1))}$	$T^3(\ln|A|)^C$	$\log|A|$	0

MAIN CODES

1. The Shannon Code. (Section 2.5)

A letter A_i, whose probability is $p(A_i)$, is given a binary word $f(A_i)$,

$$|f(A_i)| = \lceil -\log p(A_i) \rceil, \quad i = 1, \ldots, k.$$

The redundancy does not exceed 1.

2. The Huffman Code. (Section 2.5).

It is a best code for a stochastic source. The codes of two least probable letters differ by the last digit only.

3. The Cover and Lenng-Yan-Cheong Code. (Section 2.5).

The code is not prefix. The letters $A_1, \ldots, A_k$ are ordered by their probabilities. The codelength of A_i is $\lceil \log \frac{i+2}{2} \rceil$, $i = 1, \ldots, k$. The coding map is injective.

4. The Gilbert and Moore Code. (Section 2.5).

The letters are ordered by their probabilities. The codelength of A_i is $\lceil -\log p(A_i) \rceil + 1$. The redundancy does not exceed 2. The code preserves the order.

5. The Khodak Code. (Section 2.7).

It is a variable-to-block code. It is defined by a tree. Minus log probability of each leaf does not exceed a given number, whereas minus log probability of one of its sons does. The redundancy is O $\left(\frac{1}{d}\right)$, d being the coding delay.

6. The Levenstein Code. (Section 2.2).

It is a prefix encoding of integers. An integer x is given a binary word Lev x,

$$|\text{Lev } x| = \log x + \log \log x (1 + \text{O}(1)).$$

7. Empirical Combinatorial Entropic (ECE) Code. (Section 2.9).

A word w is broken into n-length subwords, $n > 0$. The list of all different subwords is the vocabulary V of w. Each subword is given its number in V. The concatenation of those numbers is the main part of ECE-code. There is a prefix, allowing one to find V and n.

8. The Shannon Empirical Code. (Section 2.9).

Subwords of a word w are encoded in accordance with their frequencies.

9. Move to Front Code. (Sections 2.9, 3.5).

The letters of an alphabet A are supposed to make a pile. The position of a letter in the pile is encoded by the monotone code. Take letters of a word w one by one. Each letter is given the code of its position in the pile. The letter is taken out of the pile and put on the top. The codelength does not exceed

$$|w|\, H_A(w) + (\log \log |A| + C) \cdot |w|,$$

$H_A(w)$ being the empirical entropy of w.

10. Empirical Entropic (EE) Code. (Section 2.8).

A word w over an alphabet B is given a vector

$$r(w) = (r_1(w), \ldots, r_{|B|}(w)),$$

$r_i(w)$ being the number of occurrences of i in w, $i = 1, \ldots, |B|$. The suffix of EE-code is the number of w within the set of words with one and the same vector $r(w)$. The prefix is the number of such a set.

11. Modified Empirical Entropic (MEE) Code. (Section 3.2).

A word w, $|w| = n$ over an alphabet B is given a code MEE (w),

$$|\text{MEE}(w)| = nH_B(w) + \frac{|B| - 1}{2} \log n + \mathrm{O}(B).$$

The code is asymptotically optimal for the set of Bernoulli sources. Its redundancy is asymptotically $\frac{|B|-1}{2} \log n$.

12. Trofimov's Code.

It is a variable-to-block code defined by a tree. The probability of a word w is averaged over all Bernoulli sources with respect to the Dirichlet measure with parameter $\frac{1}{2} = \left(\frac{1}{2}, \ldots, \frac{1}{2}\right)$. Minus log average probability of each leaf does not exceed a given number, whereas such a probability of one of its sons does. The redundancy of the Trofimov's code on any Bernoulli source is $\mathrm{O}\left(\frac{\log \log |L\Delta|}{\log |L\Delta|}\right)$, $|L\Delta|$ being the quantity of leaves of the coding tree.

13. Adaptive Code.

Given a sample w, the adaptive code f_w takes a word a to the code $f_w(a)$,

$$|f_w(a)| = \left\lceil \log E_{r(w)+1/2} p_S(a) \right\rceil,$$

where $p_S(a)$ is the probability for a word a to be generated by a source S, $E_{r(w)+1/2}$ is the average with respect to the Dirichlet measure with parameter $r(w) + 1/2$. This code is asymptotically the best.

14. The Ryabko Monotone Code.

Let Σ be a set of sources and for each $S \in \Sigma$ $p_S(A_1) \geq \ldots \geq p_S(A_k)$, where $p_S(A_i)$ is the probability for letter A_i to be generated by a source $S \in \Sigma$. The probabilities are not known. The best for Σ code f takes a letter A_i to $f(A_i)$,

$$|f(A_i)| = \left\lceil -\log\left(\frac{1}{i}\left(1 - \frac{1}{i}\right)^{i-1} \cdot 2^{-R(\Sigma)}\right)\right\rceil,$$

where

$$R(\Sigma) = \log\sum_{i=1}^{k}\frac{1}{i}\left(1 - \frac{1}{i}\right)^{i-1} + \alpha_k, \quad |\alpha_k| \leq 1.$$

$$R(\Sigma) \sim \log\log k, \quad |f(A_i)| \leq \log i + \log\log k + O(1), \quad i = 1, \ldots, k.$$

15. Rapid Lexicographic Code. (Section 5.7).

A word w is given its lexicographic number in a set S. The program length is $O(|S| \cdot |w|)$, the running time is $O(\log|S|\log|w|)$.

16. Digital Code. (Section 5.5).

It is described by a tree Δ, whose nodes are given, first, numbers from 0 to $|\Delta| - 1$, and, second, labels from 1 to $|w|$, w being a word encoded. Being at a node, take the letter at the labelled position. Go to the left or right son of the node. The word is given the number of the leaf reached. The program length is $O(|\Delta|\log|\Delta|)$, the computation time is $O(|w|)$.

17. Two-step Digital Code. (Section 5.5).

It is computed like the digital code up to a fixed level of the tree Δ. Then the numeration of the nodes is started from zero. The label of a node divides the corresponding dictionary into nearly equal parts. The program length is $O(|\Delta|\log|w|)$, the computation time is $O(|w|)$.

18. Linear Galois Code. (Section 5.2).

A word $w = w_1w_2\ldots w_{\frac{n}{\mu}}$, n being the wordlength, $\frac{n}{\mu}$ being an integer. A code φ_b is defined by an n-length word $b = b_1\ldots b_{\frac{n}{\mu}}$.

$$\varphi_b(w) = b_1w_1 + \ldots + b_{\frac{n}{\mu}}w_{\frac{n}{\mu}}.$$

19. Polynomial Galois Code. (Section 5.3).

A word $w = w_1\ldots w_{\frac{n}{\mu}}$. A code f_b is defined by an μ-length word b,

$$f_b w = w_1 + w_2 b + \ldots + w_{\frac{n}{\mu}}b^{\frac{n}{\mu}}.$$

20. Optimal Perfect Hash-code. (Section 6.3).

A word w of a dictionary S is given $\log|S|$-length code. The time to find it is either $O(\log|w|)$ or $O(\log|w|\log\log|w|)$.

REFERENCES

1. Books on Data Compression and Information Retrieval

Aho A.V., Hopcroft J.E., and Ullman J.D. (1983) Data Structures and Algorithms. Addison–Wesley, Reading, MA.

Bell T.C., Cleary J.G., Witten I.H. (1990) Text Compression. Prentice Hall, Englewood Cliffs., N.J.

Gallager R.G. (1968) Information Theory and Reliable Communication. John Wiley and Sons. Inc.

Hamming R.W. (1980) Coding. Information Theory. Prentice Hall Inc., 233 p.

Knuth D.E. (1973) The Art of Computer Programming. 2nd ed., Addison–Wesley, Reading, MA, 3 volumes.

Lynch T.J. (1985) Data Compression, Techniques and Applications. Lifetime Learning Publications, Bellmont.

Mehlhorn K. (1987) Data Structures and Algorithms. 1: Sorting and Searching Springer–Verlag.

Storer I.A. (1988) Data Compression: Methods and Theory. Computer Science Press, Rockwille, MD.

2. Books and Papers Cited

Alon N. (1986) Explicit Construction of Exponential Families of k-independent Sets. Discrete Mathematics, Vol. 58(2), P. 191–195.

Andreev A.E. (1989) On the Complexity of Partial Boolean Functions Realization by Functional Gates Networks. Discrete Mathematics, Vol. 1, P. 35–46 (Rus).

Becker B., Simon H.U. (1988) How Robust is the n-cube. Information and Computation, Vol. 77, P. 162–178.

Belichev B.F. (1963) Identifying Key of Siberian Ants. Nauka Publishing House(Rus).

Bentley J.L., Sleator D.D., Tarjan R.E., Wei V.K. (1986) A Locally Adaptive Data Compression Scheme. Comm. of ACM, Vol. 29(4), P. 320–330.

Bernstein S.N. (1946) Probability Theory. Moskow, Gostechizdat, 556p. (Rus).

Blackwell D., Girshick M.A. (1954) Theory of Games and Statistical Decisions. John Willey and Sons, Inc.

Chang C.C., Lee R.C.T. (1986) A Letter Oriented Perfect Hashing Scheme. The Computer Journal, Vol. 29(3), P. 277–281.

Cover T.M. (1973) Enumerative Source Encoding. IEEE. Trans. Inf. Th., Vol. IT-19(1), P. 73–73.

Cormac G.V., Horspool R., Kaiserwerth M. (1985) Practical Perfect Hashing. The Computer Journal, Vol. 28 (1), P. 54–58.

Davisson L. (1973) Universal Noiseless Coding. IEEE. Trans. Inf. Th., Vol. 19(6), P. 783–795.

Davisson L., Leon-Garcia A. (1980) A Source Matching Approach to Finding Minimax Codes. IEEE. Trans. Inf. Th., Vol. 26(2), P. 166–174.

Desa M. (1965) Effectiveness of Detecting or Correcting Noises. Probl. of Inf. Tr., Vol. 1(3), P. 29–39 (Rus).

Dietzfelbinger M., Karlin A., Mehlhorn K., Meyer auf der Heide F., Rohnert H., Tarjan R.E. (1988) Dynamic Perfect Hashing: Upper and Lower Bounds. 29-th Ann. Symp. on Found. of Comp. Sci., P. 524–531.

Dunham J.G. (1980) Optimal Noiseless Coding of Random Variables. IEEE. Trans. Inf. Th., Vol. IT-26(3), P. 345.

Elias P. (1975) Universal Codeword Sets and Representations of Integers. IEEE. Trans. Inf. Th., Vol.21(2), P. 194–203.

Elias P. (1987) Interval and Recency Rank Source Encoding: two on – line Adaptive Variable – Length Schemes. IEEE. Trans. Inf. Th., Vol. IT-33M, P. 3–10.

Feller W. (1957) An Introduction to Probability Theory and its Applications. John Willey and Sons, Inc., Vol. 1, second edition.

Fiat A., Naor M., Schmidt S.P., Siegel A. (1992) Nonoblivious Hashing. JACM, Vol. 39(4), P. 764–782.

Fitingof B.M. (1966) Optimal Encoding under an Unknown or Changing Statistics. Problems of Information Transmission, Vol. 2(2), P. 3–11.

Fox E.A., Chen Q., Daoud A.P. (1992) Practical Minimal Perfect Hash Function for Large Databases. Comm. of ACM, Vol. 35(1), P. 105–121.

Fox E.A., Chen Q.,Heath L. (1992) A Faster Algorithm for Constructing Minimal Perfect Hash Functions. SIGIR Forum, P. 226–273.

Fredman M., Komlos J., Szemeredi E. (1984) Storing a Sparse Table with O(1) Worst Case Access Time. JACM, Vol. 31(3), P. 538–544.

215

Friedman J. (1984) Constructing $O(n \log n)$ Size Monotone Formulae for k-th Elementary Symmetric Polynomial of n Boolean Variables. 25-th Ann. Symp. on Found. of Comp. Sci., P. 506–515.

Gilbert E.N. and Moore E.F. (1959) Variable-length Binary Encoding. BSTJ, Vol. 38(4), P. 933–967.

Goppa V.D. (1974) Arbitrary Noise Correction by Irreducible Codes. Problems of Information Transmission, Vol. 10(3), P. 118–119. (Rus).

Hansel G. (1962) Nombre des Lettres Necessaires Pour Ecrire une Fonction Symmetrique de n Variables. C.R.Acad. Sci. Paris, Vol. 261(21), P. 1297–1300.

Hartley R.V.L.(1928) Transmission of Information. BSTJ, Vol. 7(3), P. 535–563.

Huffman D.A. (1952) A Method for the Construction of Minimum-redundancy Codes. Proc. Instr. of Electr. Radio eng., Vol. 40(9), P. 1098–1101.

Jablonskii S.V. (1959) Algorithmic Difficulties of the Synthesis of Minimal Contact Networks. Problems of Cybernetics, Vol. 2, P. 75–123 (Rus).

Jacobs C.T.M., van Emde Boas P. (1986) Two Results on Tables. Inf. Proc. Lett., Vol. 22, P. 43–48.

Jaeschke G. (1981) Reciprocal Hashing – A Method for Generating minimal Perfect Hashing Functions. Comm. of ACM, Vol. 24, N 12, P. 829–833.

Jelinek F., Schneider K. (1979) On Variable-Length to Block Coding. IEEE. Trans. Inf. Th., Vol. 1(12), N. 6.

Karp R.M., Rabin M.O. (1987) Efficient Randomized Pattern-Matching Algorithms. IBM J. Res. Develop, Vol. 31(2), P. 249–260.

Kleitman D., Spencer J. (1973) Families of k-independent Sets. Discrete Mathematics, Vol. 6, P. 255–262.

Kolmogorov A.N. (1965) Three Approaches to the Definition of the Concept "the Quantity of Information". Problems of Information Transmission, Vol. 1(1), P. 3–11. (Rus).

Kolmogorov A.N., Tichomirov V.M. (1959) ε-entropy and ε-capacity of Sets in Metric Spaces. Uspechi Math. Nauk, Vol. 14(2), P. 3–86. (Rus).

Leung-Yan-Cheong S.K., Cover T.M. (1979) Some Equivalences Between Shannon Entropy and Kolmogorov Complexity. IEEE. Trans. Inf. Th., Vol. 24, N. 3, P. 331–338.

Levenstein V.I. (1968) The Redundancy and the Delay of a Decipherable Encoding of Integers. Problems of Cybernetics, Vol. 20, P. 173–179. (Rus).

Medvedev J.I. (1970) Some Theorems on Asymptotic Distribution of χ^2-statistic. Dokl. Acad. Sci. USSR, Vol. 192, N. 5, P. 987–990. (Rus).

Nagaev S.V. (1979) Large Deviations of Sums of Independent Random Variables. Ann. Probab., Vol. 7, N. 5, P. 745–789.

Nechiporuk E.I. (1965) The Complexity of Realization of Partial Boolean Functions by Valve Networks. Dokl. Acad. Sci. USSR, Vol. 163(1), P. 40–43. (Rus).

Petrov V.V. (1972) Sums of Independent Stochastic Variables. Moskow, Nauka, 414 p. (Rus).

Pippenger N. (1977) Information Theory and the Complexity of Boolean Functions. Math. Syst. Theory, Vol. 10, P. 129–167.

Poljak S., Pultr A., Rödl V. (1983) On Qualitatively Independent Partitions and Related Problems. Discr. Appl. Math., Vol. 6, P. 109–216.

Shannon C.E. (1948) A Mathematical Theory of Communication. BSTJ, Vol. 27, P. 398–403.

Shannon C.E. (1949) The Synthesis of Two-terminal Switching Circuits. BSTJ, Vol. 28, N. 1, P. 59.

Schmidt I.P., Siegel A. (1990) The Spatial Complexity of Oblivious k-probe Hash-Functions. SIAM J. Comp., Vol. 19, N. 5.

Sholomov L.A. (1969) Realization of Partial Boolean Functions by Functional Gate Networks. Problems of Cybernetics, N. 21, P. 215–227. (Rus).

Shtarkov Y., Babkin V. (1971) Combinatorial Encoding for Discrete Stationary Sources. 2-nd Int. Symp. Inf. Th., 1971, Akad. Kiado, Budapest, P. 249–257.

Sprugnoli R. (1977) Perfect Hashing Functions: a Single Probe Retrieving Method for Static Files. Comm. of ACM, Vol. 23, N. 1, P. 17–19.

Wyner A.D. (1972) An Upper Bound on the Entropy Series. Inf. Contr. Vol. 20, P. 176.

Ziv J., Lempel A. (1978) Compression of Individual Sequences via Variable–Length Coding. IEEE. Trans. Inf. Th., Vol. IT-24, N. 5, P. 530–536.

Zvonkin A.K., Levin L.A. (1970) The Complexity of Finite Objects and the Concepts of Information and Randomness through the Algorithm Theory. Uspechi Math. Nauk, Vol. 25(6), P. 85–127. (Rus).

3. The book has its origin in the following papers of the author and his students:

Hasin L.S. (1969) Complexity of Formulae over Basis $\{\vee, \&, -\}$ Realizing Threshold Functions. Dokl. Acad. Sci. USSR, Vol. 189(4), P. 752–755. (Rus).

Khodak G.L. (1969) Delay–Redundancy Relation of VB–encoding. All-union Conference on Theoretical Cybernetics. Novosibirsk. P. 12. (Rus).

Khodak G.L. (1972) Bounds of Redundancy of per Word Encoding of Bernoulli Sources. Problems of Information Transmission. Vol. 8(2), P. 21–32. (Rus).

Krichevskii R.E. (1964) π-Network Complexity of a Sequence of Boolean Functions. Problems of Cybernetics. N 12, P. 45–56. Moscow, Nauka. (Rus).

Krichevskii R.E. (1968) A Relation Between the Plausibility of Information about a Source and Encoding Redundancy. Problems of Information Transmission. Vol. 4, N 3, P. 48–57. (Rus).

Krichevskii R.E. (1976) The Complexity of Enumeration of a Finite Word Set. Dokl. Acad. Sci. USSR, Vol. 228, N 5, P. 287–290. (full version: Problems of Cybernetics N. 36, P. 159–180, (1979). English translation in Selecta Mathematica Sovetica Vol. 8, N. 2 (1989), P. 99–129).

Krichevskii R.E. (1985) Optimal Hashing. Information and Control. Vol. 62, N. 1, P. 64–92.

Krichevskii R.E. (1986) Retrieval and Data Compression Complexity. Proc. Int. Congr. Math. Berkley, 1986, P. 1461–1468.

Krichevskii R.E., Ryabko B.Ja. (1985) Universal Retrieval Trees. Discr. Appl. Math.. Vol. 12, P. 293–302.

Krichevskii R.E., Ryabko B.Ya., Haritonov A.Y. (1981) Optimal Kry for Taxons Ordered in Accordance with their Frequencies. Discr. Appl. Math.. Vol. 3, P. 67–72.

Krichevskii R.E., Trofimov V.K. (1981) The Performance of Universal Encoding. IEEE. Trans. Inf. Th.. Vol. 27, N. 2, P. 199–207.

Reznikova Zh.I., Ryabko B.Ja. (1986) Information Theoretical Analysis of the Language of Ants. Problems of Information Transmission. Vol. 22(3), P. 103–108. (Rus).

Ryabko B.Ja. (1979) The Encoding of a Source with Unknown but Ordered Probabilities. Problems of Information Transmission. Vol. 14(2), P. 71– 77. (Rus).

Ryabko B.Ja. (1980,a) Universal Encoding of Compacts. Dokl. Acad. Sci. USSR. Vol. 252(6), P. 1325–1328. (Rus).

Ryabko B.Ja. (1980,b) Information Compression by a Book Stack. Problems of Information Transmission. Vol. 16(4), P. 16–21. (Rus).

Ryabko B.Ja. (1990) A Fast Algorithm of Adaptive Encoding. Problems of Information Transmission. Vol. 26, N. 4, P. 24–37. (Rus).

Trofimov V.K. (1974) The Redundancy of Universal Encoding of Arbitrary Markov Sources. Problems of Information Transmission. Vol. 10(4), P. 16–24. (Rus).

Trofimov V.K. (1976) Universal BV-Encoding of Bernoulli Sources. Discr. Analysis, Novosibirsk, N. 29, P. 87–100.

Potapov V.N. (1993) Fast Lexicographic retrieval. Dokl. Acad. Sci. USSR. Vol. 330, N. 2, P. 158–160. (Rus).

Index

Other *Mathematics and Its Applications* titles of interest:

P.H. Sellers: *Combinatorial Complexes. A Mathematical Theory of Algorithms.*
1979, 200 pp.   ISBN 90-277-1000-7

P.M. Cohn: *Universal Algebra.* 1981, 432 pp.
                    ISBN 90-277-1213- 1 (hb), ISBN 90-277-1254-9 (pb,

J. Mockor: *Groups of Divisibility.* 1983, 192 pp.   ISBN 90-277-1539-4

A. Wwarynczyk: *Group Representations and Special Functions.* 1986, 704 pp.
                    ISBN 90-277-2294-3 (pb), ISBN 90-277-1269-7 (hb)

I. Bucur: *Selected Topics in Algebra and its Interrelations with Logic, Number
Theory and Algebraic Geometry.* 1984, 416 pp.   ISBN 90-277-1671-4

H. Walther: *Ten Applications of Graph Theory.* 1985, 264 pp.
                                                ISBN 90-277-1599-8

L. Beran: *Orthomodular Lattices. Algebraic Approach.* 1985, 416 pp.
                                                ISBN 90-277-1715-X

A. Pazman: *Foundations of Optimum Experimental Design.* 1986, 248 pp.
                                                ISBN 90-277-1865-2

K. Wagner and G. Wechsung: *Computational Complexity.* 1986, 552 pp.
                                                ISBN 90-277-2146-7

A.N. Philippou, G.E. Bergum and A.F. Horodam (eds.): *Fibonacci Numbers and
Their Applications.* 1986, 328 pp.   ISBN 90-277-2234-X

C. Nastasescu and F. van Oystaeyen: *Dimensions of Ring Theory.* 1987, 372 pp.
                                                ISBN 90-277-2461-X

Shang-Ching Chou: *Mechanical Geometry Theorem Proving.* 1987, 376 pp.
                                                ISBN 90-277-2650-7

D. Przeworska-Rolewicz: *Algebraic Analysis.* 1988, 640 pp.   ISBN 90-277-2443-1

C.T.J. Dodson: *Categories, Bundles and Spacetime Topology.* 1988, 264 pp.
                                                ISBN 90-277-2771-6

V.D. Goppa: *Geometry and Codes.* 1988, 168 pp.   ISBN 90-277-2776-7

A.A. Markov and N.M. Nagorny: *The Theory of Algorithms.* 1988, 396 pp.
                                                ISBN 90-277-2773-2

E. Kratzel: *Lattice Points.* 1989, 322 pp.   ISBN 90-277-2733-3

A.M.W. Glass and W.Ch. Holland (eds.): *Lattice-Ordered Groups. Advances and
Techniques.* 1989, 400 pp.   ISBN 0-7923-0116-1

N.E. Hurt: *Phase Retrieval and Zero Crossings: Mathematical Methods in Image
Reconstruction.* 1989, 320 pp.   ISBN 0-7923-0210-9

Du Dingzhu and Hu Guoding (eds.): *Combinatorics, Computing and Complexity.*
1989, 248 pp.   ISBN 0-7923-0308-3

Other *Mathematics and Its Applications* titles of interest:

A.Ya. Helemskii: *The Homology of Banach and Topological Algebras.* 1989, 356 pp.   ISBN 0-7923-0217-6

J. Martinez (ed.): *Ordered Algebraic Structures.* 1989, 304 pp.
                                                                                                ISBN 0-7923-0489-6

V.I. Varshavsky: *Self-Timed Control of Concurrent Processes. The Design of Aperiodic Logical Circuits in Computers and Discrete Systems.* 1989, 428 pp.
                                                                                                ISBN 0-7923-0525-6

E. Goles and S. Martinez: *Neural and Automata Networks. Dynamical Behavior and Applications.* 1990, 264 pp.   ISBN 0-7923-0632-5

A. Crumeyrolle: *Orthogonal and Symplectic Clifford Algebras. Spinor Structures.* 1990, 364 pp.   ISBN 0-7923-0541-8

S. Albeverio, Ph. Blanchard and D. Testard (eds.): *Stochastics, Algebra and Analysis in Classical and Quantum Dynamics.* 1990, 264 pp.   ISBN 0-7923-0637-6

G. Karpilovsky: *Symmetric and G-Algebras. With Applications to Group Representations.* 1990, 384 pp.   ISBN 0-7923-0761-5

J. Bosak: *Decomposition of Graphs.* 1990, 268 pp.   ISBN 0-7923-0747-X

J. Adamek and V. Trnkova: *Automata and Algebras in Categories.* 1990, 488 pp.
                                                                                                ISBN 0-7923-0010-6

A.B. Venkov: *Spectral Theory of Automorphic Functions and Its Applications.* 1991, 280 pp.   ISBN 0-7923-0487-X

M.A. Tsfasman and S.G. Vladuts: *Algebraic Geometric Codes.* 1991, 668 pp.
                                                                                                ISBN 0-7923-0727-5

H.J. Voss: *Cycles and Bridges in Graphs.* 1991, 288 pp.   ISBN 0-7923-0899-9

V.K. Kharchenko: *Automorphisms and Derivations of Associative Rings.* 1991, 386 pp.   ISBN 0-7923-1382-8

A.Yu. Olshanskii: *Geometry of Defining Relations in Groups.* 1991, 513 pp.
                                                                                                ISBN 0-7923-1394-1

F. Brackx and D. Constales: *Computer Algebra with LISP and REDUCE. An Introduction to Computer-Aided Pure Mathematics.* 1992, 286 pp.
                                                                                                ISBN 0-7923-1441-7

N.M. Korobov: *Exponential Sums and their Applications.* 1992, 210 pp.
                                                                                                ISBN 0-7923-1647-9

D.G. Skordev: *Computability in Combinatory Spaces. An Algebraic Generalization of Abstract First Order Computability.* 1992, 320 pp.   ISBN 0-7923-1576-6

E. Goles and S. Martinez: *Statistical Physics, Automata Networks and Dynamical Systems.* 1992, 208 pp.   ISBN 0-7923-1595-2

Other *Mathematics and Its Applications* titles of interest:

M.A. Frumkin: *Systolic Computations.* 1992, 320 pp.   ISBN 0-7923-1708-4

J. Alajbegovic and J. Mockor: *Approximation Theorems in Commutative Algebra.* 1992, 330 pp.   ISBN 0-7923-1948-6

I.A. Faradzev, A.A. Ivanov, M.M. Klin and A.J. Woldar: *Investigations in Algebraic Theory of Combinatorial Objects.* 1993, 516 pp.   ISBN 0-7923-1927-3

I.E. Shparlinski: *Computational and Algorithmic Problems in Finite Fields.* 1992, 266 pp.   ISBN 0-7923-2057-3

P. Feinsilver and R. Schott: *Algebraic Structures and Operator Calculus. Vol. 1. Representations and Probability Theory.* 1993, 224 pp.   ISBN 0-7923-2116-2

A.G. Pinus: *Boolean Constructions in Universal Algebras.* 1993, 350 pp.
ISBN 0-7923-2117-0

V.V. Alexandrov and N.D. Gorsky: *Image Representation and Processing. A Recursive Approach.* 1993, 200 pp.   ISBN 0-7923-2136-7

L.A. Bokut' and G.P. Kukin: *Algorithmic and Combinatorial Algebra.* 1993, 469 pp.   ISBN 0-7923-2313-0

Y. Bahturin: *Basic Structures of Modern Algebra.* 1993, 419 pp.
ISBN 0-7923-2459-5

R. Krichevsky: *Universal Compression and Retrieval.* 1994, 219 pp.
ISBN 0-7923-2672-5